Studies in Fuzziness and Soft Computing

Volume 310

For further volumes:
http://www.springer.com/series/2941

About this Series

The series "Studies in Fuzziness and Soft Computing" contains publications on various topics in the area of soft computing, which include fuzzy sets, rough sets, neural networks, evolutionary computation, probabilistic and evidential reasoning, multi-valued logic, and related fields. The publications within "Studies in Fuzziness and Soft Computing" are primarily monographs and edited volumes. They cover significant recent developments in the field, both of a foundational and applicable character. An important feature of the series is its short publication time and world-wide distribution. This permits a rapid and broad dissemination of research results.

Vicenç Torra · Yasuo Narukawa · Michio Sugeno
Editors

Non-Additive Measures

Theory and Applications

Editors
Vicenç Torra
IIIA-CSIC
Bellaterra
Catalonia
Spain

Michio Sugeno
European Centre for Soft Computing
Mieres
Spain

Yasuo Narukawa
Toho Gakuen
Kunitachi
Tokyo
Japan

ISSN 1434-9922 ISSN 1860-0808 (electronic)
ISBN 978-3-319-37525-0 ISBN 978-3-319-03155-2 (eBook)
DOI 10.1007/978-3-319-03155-2
Springer Cham Heidelberg New York Dordrecht London

Printed on acid-free paper

Springer is part of Springer Science+Business Media (www.springer.com)

Preface

A measure assigns values to sets and generalizes the concept of length, area and volume. Probability measures are a well known example of measures. Lebesgue measures are another example. Most typically, measures are additive, as length, areas and volumes are. That is, the measure of the union of two disjoint sets is the sum of these two measures

$$\mu(A \cup B) = \mu(A) + \mu(B)$$

for $A \cap B = \emptyset$. Again, probabilities are an example of additive measures.

Although not so much known, non-additive measures have also been studied in the literature both for their mathematical properties as well as for their application to real problems. Non-additive measures replace additivity my monotonicity. That is,

$$\mu(A) \leq \mu(C) \text{ if } A \subseteq C.$$

As all additive measures are monotonic, non-additive measures generalize additive ones.

Non-additive measures are also known by the term capacities, and fuzzy measures.

Non-additive measures permit us to represent interaction between the elements. For example, we might have $\mu(A \cup B) < \mu(A) + \mu(B)$ (negative interaction between A and B), and $\mu(A \cup B) > \mu(A) + \mu(B)$ (positive interaction between A and B).

Then, in the same way that there are fundamental concepts in measure theory based on additive measures, there are some based on non-additive measures. Some of these concepts are generalizations of corresponding concepts for additive measures. For example, the Choquet integral [1] is a generalization of the Lebesgue integral in the sense that the Choquet integral of a function f with respect to an additive measure corresponds to the Lebesgue integral. Other concepts were introduced as new in this non-additive setting. This is the case of the integral introduced by Sugeno in 1972 [4,5] which is now known as the Sugeno integral.

This book has its origin in the 9th International Conference on Modeling Decisions for Artificial Intelligence (MDAI 2012) that took place in Girona[1] and, more specifically, in the panel session *Fuzzy measures, fuzzy integrals and aggregation operators* hold in the conference. The panel gathered key researchers with the aim of discussing new and challenging lines for future research in the area of non-additive measures and integrals. The chapters of this book, written by most of the panelists and two additional invited authors, are state-of-the-art descriptions of the field that cover the lines of research discussed in the panel.

The first chapter is a review of uses and applications of non-additive measures and integrals. The chapter presents most relevant definitions and also points out to the other chapters in the book for further details and references. Links between non-additive integrals and aggregation operators [6] are also highlighted.

In the second chapter, Narukawa presents an overview of integration with respect to a non-additive measure. The chapter gives special emphasis to integrals over continuous domains. The Sugeno, Choquet and generalized integrals are presented and their properties reviewed. The case of multidimensional integrals are discussed and a Fubini-like theorem is presented. The chapter concludes with the Möbius transform and generalizations of the Möbius transform.

In the third chapter, Mesiar and Stupňanová focuses on different integrals with respect to a non-additive measure. The authors discuss the approach to integration introduced by Even and Lehrer [2] (decomposition integrals), the Choquet and Sugeno integrals and also Shilkret and universal integrals.

Chapter four, by Honda, focuses on the definition of entropy for non-additive measures (or capacities). First, Honda reviews the definition of entropy for probabilities and then introduces different generalizations that exist for non-additive measures. The author not only considers the case of measures defined on 2^X but also on measures defined on set systems (based on a subset of 2^X). The problem of the axiomatization of entropies is also discussed.

Non-additive measures and integrals have been used in applications. The fifth chapter by Ozaki focuses on their application to economics. More especifically, the author considers decision theory under risk, and decision theory under uncertainty. The chapter describes some of the problems and paradoxes that cannot be solved using additive models (as e.g. Ellsberg's paradox).

Fujimoto in Chapter six surveys cooperative game theory, an important application area for non-additive measures. Non-additive measures permit to represent coalitions in game theory. The chapter discusses with detail the case in which not all coalitions can be formed, and how we can deal with this situation. The chapter also discusses indices that have been defined for games (as the Shapley index [3]).

Flaminio, Godo, and Kroupa focus in Chapter seven on belief functions on MV-algebras of fuzzy sets. Belief functions are totally monotone non-additive measures. The authors discuss two ways of extending belief functions on Boolean algebras of events to MV-algebras of events.

[1] http://www.mdai.cat/mdai2012

We hope that this book will provide a reference to students, researchers and practitioners in the field.

The editors of this book would like to thank Prof. Kacprzyk for his encouragement to edit it and publish it in this series. Partial support by the Spanish MINECO (TIN2011-15580-E) is acknowledged.

August 2013
L'Alcalatén

Vicenç Torra
Yasuo Narukawa
Michio Sugeno

References

1. Choquet, G. (1953/54) Theory of capacities, Ann. Inst. Fourier 5 131-295.
2. Even, Y., Lehrer, E. (2013) Decomposition-Integral: Unifying Choquet and the Concave Integrals, Submitted.
   http://www.tau.ac.il/~ lehrer/Papers/decomposition.pdf (access August 2013)
3. Shapley, L. (1953) A value for n-person games, Annals of Mathematical Studies 28 307-317.
4. Sugeno, M. (1972) Fuzzy measures and fuzzy integrals (in Japanese), Trans. of the Soc. of Instrument and Control Engineers 8:2
5. Sugeno, M. (1974) Theory of Fuzzy Integrals and its Applications, Ph. D. Dissertation, Tokyo Institute of Technology, Tokyo, Japan.
6. Torra, V., Narukawa, Y. (2007) Modeling decisions: information fusion and aggregation operators, Springer.

Contents

Use and Applications of Non-Additive Measures and Integrals

Vicenç Torra

IIIA, Institut d'Investigació en Intel·ligència artificial,
CSIC, Consejo Superior de Investigaciones Científicas,
Campus Universitat Autònoma de Barcelona s/n,
08193 Bellaterra, Catalonia
vtorra@iiia.csic.es, vtorra@ieee.org

Abstract. Non-additive measures (also known as fuzzy measures and capacities) and integrals have been used in several types of applications. In this chapter we review the main definitions related to these measures, motivate their use from the point of view of the applications, and describe their use in different contexts.

1 Introduction

Non-additive measures and integrals have been used in different areas. They are used to overcome the shortcommings of other (simpler) models. In particular, they are used when models based on additive measures are not appropriate. In this chapter we review the basic definitions of this topic and describe some of the problems that can be solved using non-additive measures.

For the sake of completeness, Section 2 reviews definitions and results that are needed in the rest of the chapter. Then, Section 3 focuses on decision making. We describe some paradoxes that can be solved using non-additive measures and integrals and that cannot be solved with some of the alternative models. Section 4 focuses on the use of these integrals on subjective evaluation. Section 5 reviews some other applications in e.g. the field of computer vision. The chapter finishes with some conclusions.

In this chapter we also establish some links between the topics described here and the other chapters of this book.

2 Preliminaries

This section reviews a few definitions on probability, measures, aggregation operators and integrals. We start reviewing the definition of σ-algebra. The notion of σ-algebra is introduced because for some sets X it is, in general, not possible to consider all subsets of X and define the probability for these subsets. In particular, generally it is not possible when X is not finite. A σ-algebra is a set of sets with appropriate properties to define the measure on them. See [26] for details.

V. Torra, Y. Narukawa, and M. Sugeno (eds.), *Non-Additive Measures*,
Studies in Fuzziness and Soft Computing 310,
DOI: 10.1007/978-3-319-03155-2_1,

Definition 1. *Let X be a reference set, and let $\mathcal{A}$ be a subset of $\wp(X)$. Let us consider the following properties:*

Property 1: $\emptyset \in \mathcal{A}$ *and* $X \in \mathcal{A}$
Property 2: *if* $A \in \mathcal{A}$ *then* $X \setminus A \in \mathcal{A}$
Property 3: $\mathcal{A}$ *is closed under* finite *unions and* finite *intersections:*

$$\text{if } A_1, \ldots, A_n \in \mathcal{A}, \text{ then } \cup_{i=1}^{n} A_i \in \mathcal{A} \text{ and } \cap_{i=1}^{n} A_i \in \mathcal{A},$$

Property 4: $\mathcal{A}$ *is closed under* countable *unions and intersections:*

$$\text{if } A_1, A_2, \cdots \in \mathcal{A}, \text{ then } \cup_{i=1}^{\infty} A_i \in \mathcal{A} \text{ and } \cap_{i=1}^{\infty} A_i \in \mathcal{A},$$

Then, we define algebra and σ-algebra as follows.

1. *$\mathcal{A}$ is an algebra (over a set) (or a field) if it satisfies Properties 1, 2, and 3;*
2. *$\mathcal{A}$ is a σ-algebra (or a σ-field) if it satisfies Properties 1, 2, and 4.*

Note that Properties 1 and 4 imply Property 3. Therefore, any σ-algebra is an algebra. Nevertheless, Properties 1 and 3 do not imply Property 4.

When $\mathcal{A}$ is an algebra on the reference set X, $\mathcal{A}$ is called a measurable space, the elements of $\mathcal{A}$ are called measurable sets, and a pair $(X, \mathcal{A})$ is called a measurable space.

In probability, X is called the *outcome space* (and *sample space* in statistics), elements in X are the possible *outcomes*, sets $A \in \mathcal{A}$ are called *events*, and a pair $(X, \mathcal{A})$ is also called a measurable space.

When X is finite, $\wp(X)$ is a σ-algebra and, therefore, $(X, \wp(X))$ is a measurable space.

When X is the set of real numbers, it is usual to consider the Borel σ-algebra. We define it below. We begin defining the σ-algebra generated by a class of subsets $\mathcal{S}$ of X. This σ-algebra is denoted by $\sigma(\mathcal{S})$.

Definition 2. *Let $\mathcal{S}$ be a class of subsets of X; then, the σ-algebra generated by $\mathcal{S}$ is the set of subsets of X that*

1. *contains $\mathcal{S}$*
2. *is a σ-algebra*
3. *is as small as possible (in the sense that any other set of subsets that contains $\mathcal{S}$ also contains $\sigma(\mathcal{S})$).*

Let $\mathcal{O}$ be the set of all finite open subsets of $\mathbb{R}$; then, $\mathcal{B} = \sigma(\mathcal{O})$ is the Borel σ-algebra of $\mathbb{R}$.

2.1 Measures and Probability Measures

Let us start defining measures. See [26,66] for details.

Definition 3. *Let $(X, \mathcal{A})$ be a measurable space; then, a set function μ is an additive measure if it satisfies the following conditions:*

(i) *$\mu(A) \geq 0$ for all $A \in \mathcal{A}$,*
(ii) *$\mu(X) \leq \infty$*
(iii) *$\mu(\cup_{i=1}^{\infty} A_i) = \sum_{i=1}^{\infty} \mu(A_i)$ for every countable sequence A_i $(i \geq 1)$ of $\mathcal{A}$ that is pairwise disjoint (i.e,. $A_i \cap A_j = \emptyset$ when $i \neq j$).*

The triple $(X, \mathcal{A}, \mu)$ is known as a measure space.

Probability is an additive measure such that $\mu(X) = 1$. We usually denote it with P. The triple $(X, \mathcal{A}, P)$ is known as a probability space.

Example 1. There is a unique measure m on $(\mathbb{R}, \mathcal{B})$ that satisfies $m([a, b]) = b - a$ for every finite $[a, b]$ with $-\infty < a \leq b < \infty$. It is the Lebesgue measure.

Definition 4. *Let $(X, \mathcal{A})$ and $(Y, \mathcal{B})$ be two measurable spaces and f a function from X to Y. Then, it is said that the function f is a measurable function from $(X, \mathcal{A})$ to $(Y, \mathcal{B})$ if and only if $f^{-1}(B) \in \mathcal{A}$. Here, $f^{-1}(B)$ is defined by $\{x \in X | f(x) \in B\}$.*

A random variable is a function that assigns values in $\mathbb{R}$ to the outcomes in the reference set.

Definition 5. *Let $(X, \mathcal{A})$ and $(\mathbb{R}, \mathcal{B})$ be two measurable spaces, then a measurable function f from X to $\mathbb{R}$ is a random variable.*

Given a probability space $(X, \mathcal{A}, P)$ and a random variable $f : X \to S \subseteq \mathbb{R}$, we can define a new probability measure on the space S from P. This is shown below. Note that in this definition we need a σ-algebra $\mathcal{S}$ on the space S, as the new probability should be defined over subsets of S.

Definition 6. *Let $(X, \mathcal{A}, P)$ be a probability space and let $f : X \to S$ be a random variable. Then, the expectation of f is defined by*

$$E[f] := \int_X f dP$$

provided the integral exists.

2.2 Non-Additive Measures

In this section we review the definition of non-additive measures, also known as fuzzy measures and capacities, and some families of these measures.

Definition 7. *Let $(X, \mathcal{A})$ be a measurable space, a non-additive (fuzzy) measure μ on $(X, \mathcal{A})$ is a set function $\mu : \mathcal{A} \to [0, 1]$ satisfying the following axioms:*

(i) $\mu(\emptyset) = 0$, $\mu(X) = 1$ (boundary conditions)
(ii) $A \subseteq B$ implies $\mu(A) \leq \mu(B)$ (monotonicity)

Note that a probability is a fuzzy measure, as the condition of additivity implies monotonicity. Nevertheless, it is not true on the other way round. That is, monotonicity does not imply additivity.

Let us now consider some definitions related to these measures.

Definition 8. *Let μ be a non-additive measure on the measurable space $(X, \mathcal{A})$. Then,*

- *μ is additive if $\mu(A \cup B) = \mu(A) + \mu(B)$ when $A \cap B = \emptyset$;*
- *μ is null-additive if $A, B \in \mathcal{A}$ with $A \cap B = \emptyset$, $\mu(B) = 0$ implies $\mu(A \cup B) = \mu(A)$;*
- *μ is submodular if $\mu(A) + \mu(B) \geq \mu(A \cup B) + \mu(A \cap B)$;*
- *μ is supermodular if $\mu(A) + \mu(B) \leq \mu(A \cup B) + \mu(A \cap B)$.*
- *μ is symmetric if for finite X, when $|A| = |B|$, then $\mu(A) = \mu(B)$. In general if $X \subseteq \mathbb{R}$, on μ is said to be symmetric if $m(A) = m(B)$ implies $\mu(A) = \mu(B)$ (where m represents the Lebesgue measure).*

Non-additive measures can be represented through the Möbius transform. The Möbius transform of a non-additive measure μ on a finite set X is the function $m : \wp(X) \to \mathbb{R}$ defined below. The case of the Möbius transform of a fuzzy measure on infinite sets is described in [49] (See Section 5 of Chapter 2 in this book).

Definition 9. *Let μ be a non-additive measure on a finite set X; then, its Möbius transform m is defined as*

$$m_\mu(A) := \sum_{B \subseteq A} (-1)^{|A|-|B|} \mu(B) \tag{1}$$

for all $A \subset X$.

The set of subsets A of X such that $m_\mu(A)$ is not zero defines the focal set. That is, $\mathcal{F}_\mu = \{A | m_\mu(A) \neq 0\}$. If A is in $\mathcal{F}_\mu$ we say that A is a focal element.

Note that the function m is not restricted to the $[0, 1]$ interval, but that $m(\emptyset) = 0$, that $\sum_{A \subseteq X} m(A) = 1$, and, if $A \subset B$, then $\sum_{C \subseteq A} m(C) \leq \sum_{C \subseteq B} m(C)$.

Given a function m that is a Möbius transform (i.e., that satisfies the three conditions in the last paragraph), we can reconstruct the original measure as follows:

$$\mu(A) = \sum_{B \subseteq A} m(B)$$

for all $A \subseteq X$.

There are several families of measures. We review some of them below. There are several motivations for studying these families. One of them is because it is more complex to define a non-additive measure than an additive one because we need to assign a number to each set in the σ-algebra. In the case of a finite set X, this implies that we need to assign values to all $A \in \wp(X)$ which are $2^{|X|}$ values. On the contrary, for additive measures only $|X|$ values are needed.

Decomposable Measures. Some families are defined in a way that the measure of a set can be computed from the values in the singletons and a parameter that describes how the values of the singletons have to be combined. One way to combine the values in the singletons is using a t-conorm.

Definition 10. *Let μ be a non-additive measure on a finite set X; then, μ is a* $\perp$-decomposable measure *if there exists a t-conorm $\perp$ such that,*

$$\mu(A \cup B) = \mu(A) \perp \mu(B) \tag{2}$$

for all $A \cap B = \emptyset$

When $\sum_{x \in X} \mu(\{x\}) = 1$ and $\perp$ is the Lukasiewicz t-conorm ($\perp(x, y) = min(x + y, 1)$), the previous definition corresponds to a probability measure. Sugeno λ-measures are an example of the previous family with Sugeno's t-conorm $\perp(x, y) = min(1, x + y - \lambda xy)$.

Definition 11. *Let μ be a non-additive measure on a finite set X; then, μ is a* Sugeno λ-measure *if for some fixed $\lambda > -1$ it holds that*

$$\mu(A \cup B) = \mu(A) + \mu(B) + \lambda\mu(A)\mu(B) \tag{3}$$

for all $A \cap B = \emptyset$

Naturally, when $\lambda = 0$, we have that the Sugeno λ-measure is a probability distribution if $\mu(X) = 1$, and when $\lambda = -1$, we have that $\mu(A \cup B) = S(\mu(A), \mu(B))$ for the t-conorm $S(a, b) = a + b - ab$ (the algebraic sum).

When we require that $\mu(X) = 1$, the measures on the singletons determine the parameter λ. Note that for finite X we have

$$\mu(X) = (1/\lambda)(\Pi_{x_i \in X}[1 + \lambda v(x_i)] - 1) = 1.$$

The following proposition establishes this result.

Proposition 1. *[41] Let μ be a Sugeno λ-measure on a finite reference set $X = \{x_1, \ldots, x_N\}$; then, for a fixed set of $0 < \mu(\{x_i\}) < 1$, there exists a unique $\lambda \in (-1, +\infty)$ and $\lambda \neq 0$ that satisfies $\mu(X) = 1$, that is, satisfies*

$$\lambda + 1 = \Pi_{i=1}^{N}(1 + \lambda\mu(\{x_i\})).$$

We consider now the family of distorted probabilities.

Definition 12. *Let f be a real-valued function on $[0, 1]$ and let P be a probability measure on $(X, \wp(X))$ with finite X. We say that f and P represent a fuzzy measure μ on $(X, \wp(X))$ if and only if $\mu(A) = f(P(A))$ for all $A \in \wp(X)$.*

Definition 13. *Let f be a real-valued function on $[0, 1]$. We say that f is strictly increasing with respect to a probability measure P if and only if $P(A) < P(B)$ implies $f(P(A)) < f(P(B))$. We say that f is nondecreasing with respect to a probability measure P if and only if $P(A) < P(B)$ implies $f(P(A)) \leq f(P(B))$.*

Using the two definitions, we define distorted probabilities as follows.

Definition 14. *Let μ be a non-additive measure on $(X, \wp(X))$ with a finite X. We say that μ is a distorted probability if it is represented by a probability distribution P on $(X, \wp(X))$ and a function f that is nondecreasing with respect to a probability P.*

Proposition 2. *(see e.g. [86] for a proof) Any Sugeno λ-measure is a distorted probability.*

The family of Hierarchically $\perp$-Decomposable Fuzzy Measures (HDFM) is another family of non-additive measures. They are a generalization of Sugeno λ-measures. The difference is that in the Sugeno λ-measure the measure of any aggregation of sets is always computed using the same expression (i.e., Equation 3 with a given λ) and this is independent of the sets under consideration. In contrast, in a HDFM the combination depends on the sets. In other words, in a HDFM given disjoints A and B, $\mu(A \cup B)$ is a function of $\mu(A)$ and $\mu(B)$ which depend on A and B. For example, we can have something like:

$$\mu(A \cup B) = \mu(A) + \mu(B) + \lambda(A, B)\mu(A)\mu(B).$$

In the general expression, a t-conorm is used to combine $\mu(A)$ and $\mu(B)$.

In order to know which t-conorm is used for a particular pair of sets, we start with a dendrogram of the reference set. The leaves of the dendrogram are labeled with the measures of the singletons and the remaining nodes are labeled with t-conorms. Then, given a particular set A, A is decomposed in a hierarchical way according to the dendrogram. This dendogram will typically have less nodes, and only the leaves in the set A. The values of the leaves are then combined using the t-conorms in the nodes.

The formalization of HDFM is based first on the hierarchy of elements. Its definition follows. The original definition of these measures can be found in [77], and a more detailed explanation with examples can be found in [86].

Definition 15. *H is a hierarchy of a finite set of elements X if and only if the following conditions are fulfilled:*

(i) All the elements in X belong to the hierarchy, and they define the leaves of the hierarchy:
For all x in X, $\{x\} \in H$.

(ii) There is only one root in the hierarchy, and it is denoted by root. *A node is the root if it is not included in any other node:*
if $\text{root} \in H$, *then there is no other node $m \in H$ such that* $\text{root} \in m$.

(iii) All nodes belong to one and only one node, except for the root:
if $n \in H$ and $n \neq$ root, *then there exists a single $m \in H$ such that $n \in m$.*

(iv) All nodes that contain only one element are singletons:
if $|n| = 1$, then there exists $x \in X$ such that $n = \{x\}$ for all $n \in H$.

(v) All non-singletons are defined in terms of nodes that are in the tree:
if $|n| \neq 1$, then, for all $n_i \in n$, $n_i \in H$.

As stated before, given a dendrogram, we label the leaves with values and the other nodes with t-conorms. This is shown in the next definition.

Definition 16. *Let H be a hierarchy according to Definition 15; then, a* labeled hierarchy *L for H is a tuple $L = < H, \perp, m >$, where $\perp$ is a function that maps each node $n \in H$ that is not a leaf into a t-conorm, and m is a function that maps each singleton into a value of the unit interval.*

For simplicity, we will express $\perp(h)$ by $\perp_h$.

To complete the definition of HDFM we need to define *extension of a node.* It is the set of elements of X which are leaves of the node.

Definition 17. *Let H be a hierarchy according to Definition 15 and let h be a node in H; then, the* extension *of h in H is defined as:*

$$EXT(h) := \begin{cases} h & \text{if } |h| = 1 \\ \cup_{h_i \in h} EXT(h_i) & \text{if } |h| \neq 1. \end{cases}$$

Now, we review this family of measures.

Definition 18. *Let $L =< H, \bot, m >$ be a labeled hierarchy according Definition 16; then, the corresponding* Hierarchically $\bot$-Decomposable Fuzzy Measure *(HDFM for short) of a set B is defined as $\mu(B) = \mu_{root}(B)$, where μ_A for a node $A = \{a_1, ..., a_n\}$ is defined recursively as*

$$\mu_A(B) = \begin{cases} 0 & \text{if } |B| = 0 \\ m(B) & \text{if } |B| = 1 \\ \bot_A(\mu_{a_1}(B_1), ..., \mu_{a_n}(B_n)) & \text{if } |B| > 1. \end{cases}$$

Here, $B_i = B \cap EXT(a_i)$ for all a_i in A.

The families of measures seen in this section have been mainly defined and studied for finite X.

Level-Dependent Fuzzy Measures. Greco, Matarazzo and Giove [25] introduced a generalization of the Choquet integral considering a set of fuzzy measures. In this set, measures are indexed by a parameter t that corresponds to the *level* in the integration process. We review these measures below. The level-dependent Choquet integral is introduced in Definition 35.

Definition 19. *[25] Let $(X, \mathcal{A})$ be a measurable space. A level-dependent fuzzy measure is a function $\mu : \mathcal{A} \times (\alpha, \beta) \to [0, 1]$, $(\alpha, \beta) \subseteq \mathbb{R}$, such that*

1. *for all $t \in (\alpha, \beta)$ and $A \subseteq B$ with $A, B \in \mathcal{A}$, $\mu(A, t) \leq \mu(B, t)$,*
2. *for all $t \in (\alpha, \beta)$, $\mu(\emptyset, t) = 0$ and $\mu(X, t) = 1$,*
3. *for all $A \in \mathcal{A}$, $\mu(A, t)$ considered as a function with respect to t is Lebesgue mesurable.*

Note that for a given t, $\mu^t(x, t)$ is a fuzzy measure.

Other Extensions. Non-additive measures correspond to cooperative games in game theory. Then, a non-additive measure represents the value of a coalitions. Formally, we have that $\mu(A)$ is the value of a given coalition A. For example, we can represent the case that X is the set of parties in a parliament and $\mu(A) \in \{0, 1\}$ represents whether the coalition $A \subseteq X$ is able to approve a law.

In the setting of games it is of relevance to study the case where some coalitions are not possible because they cannot be formed. E.g., parties x_1 and x_2 cannot collaborate in approving the law.

This situation is described in detail in [18] (Chapter 6 in this book). See also the book by Bilbao [6].

Another type of extension is when the measure is not defined on a set but on a multiset. That is, we allow multiple appearances of the same object when computing the fuzzy measure. This situation has been studied in e.g. [53,90].

2.3 Aggregation Operators

In this section we focus on the approaches to aggregate data supplied by a finite set $X = \{x_1, \ldots, x_N\}$ of objects. We consider the aggregation of data proceeding from the elements of this set. Typically, X represents the data sources that supply some information to be aggregated. For example, X can be a set of experts giving opinions, a set of criteria used to evaluate some alternatives, and a set of sensors. Then, we will consider that we have a function that relates each data source with the value its supplies. Let f be such function that assigns a value in a certaine range $\mathbb{D}$ to each x in X. That is, $f : X \to \mathbb{D}$. We will use a_i to denote $f(x_i)$. Then, aggregation operators take N values $a_1, \ldots, a_N$ and combine them in a single datum. That is, $\mathbb{C} : \mathbb{D}^N \to \mathbb{D}$. $\mathbb{C}$ is usually required to satisfy idempotency (unanimity) and monotonicity. See [86,5] for details on aggregation operators. [86] gives examples of their application and discusses their differences. In the remaining part of this section we review the most rellevant functions.

Definition 20. *Given a domain $\mathbb{D}$ with a total order $\leq_{\mathbb{D}}$, an aggregation operator is a function $\mathbb{C} : \mathbb{D}^N \to \mathbb{D}$ satisfying idempotency and monotonicity.*

In this section we review some of these operators, mainly focusing to the case where $\mathbb{D}$ is a subset of $\mathbb{R}$.

Definition 21. *A vector $v = (v_1 ... v_N)$ is a* weighting vector *of dimension N if and only if $v_i \in [0, 1]$ and $\sum_i v_i = 1$.*

Definition 22. *A mapping AM: $\mathbb{R}^N \to \mathbb{R}$ is an* arithmetic mean *of dimension N if $AM(a_1, ..., a_N) = (1/N) \sum_{i=1}^{N} a_i$.*

Definition 23. *Let $\mathbf{p}$ be a weighting vector of dimension N; then, a mapping WM: $\mathbb{R}^N \to \mathbb{R}$ is a* weighted mean *of dimension N if $WM_{\mathbf{p}}(a_1, ..., a_N) = \sum_{i=1}^{N} p_i a_i$.*

Note that the expectation as defined in Definition 6 is the continuous counterpart of the expression given here for the weighted mean. The weighting vector $\mathbf{p}$ corresponds to the probability on X.

Definition 24. *Let* $\mathbf{w}$ *be a weighting vector of dimension* N*; then, a mapping* $OWA: \mathbb{R}^N \to \mathbb{R}$ *is an* Ordered Weighting Averaging (OWA) operator *of dimension* N *if*

$$OWA_{\mathbf{w}}(a_1, ..., a_N) = \sum_{i=1}^{N} w_i a_{\sigma(i)},$$

where $\{\sigma(1), ..., \sigma(N)\}$ *is a permutation of* $\{1, ..., N\}$ *such that* $a_{\sigma(i-1)} \geq a_{\sigma(i)}$ *for all* $i = \{2, ..., N\}$ *(i.e.,* $a_{\sigma(i)}$ *is the* i*th largest element in the collection* $a_1, ..., a_N$*).*

Both OWA and weighted mean generalize the arithmetic mean.

The WOWA Operator. The WOWA operator was introduced in order to aggregate data taking into account the weights of both the weighted mean and the OWA. That is, to model situations in which both importance of information sources (as in the weighted mean) and importance of values (as in the OWA) had to be taken into account. The definition of the operator is as follows. See [82] for a review of results related to the WOWA operator.

Definition 25. *[74,75] Let* $\mathbf{p}$ *and* $\mathbf{w}$ *be two weighting vectors of dimension* N*; then, a mapping* $WOWA: \mathbb{R}^N \to \mathbb{R}$ *is a* Weighted Ordered Weighted Averaging (WOWA) operator *of dimension* N *if*

$$WOWA_{\mathbf{p},\mathbf{w}}(a_1, ..., a_N) = \sum_{i=1}^{N} \omega_i a_{\sigma(i)},$$

where σ *is defined as in the case of OWA (*i.e., $a_{\sigma(i)}$ *is the* i*th largest element in the collection* $a_1, ..., a_N$*), and the weight* ω_i *is defined as*

$$\omega_i = w^*(\sum_{j \leq i} p_{\sigma(j)}) - w^*(\sum_{j < i} p_{\sigma(j)}),$$

with w^* *being a nondecreasing function that interpolates the points*

$$\{(i/N, \sum_{j \leq i} w_j)\}_{i=1,...,N} \cup \{(0,0)\}.$$

The function w^* *is required to be a straight line when the points can be interpolated in this way.*

The previous definition starts with the weights w and p, and from w we build the function W^*. An alternative approach is to start directly with the function W^*. This is defined below with the help of the set of functions that can be used as W^*.

Definition 26. *A function* $Q : [0,1] \to [0,1]$ *is a* regular nondecreasing fuzzy quantifier *(nondecreasing fuzzy quantifier for short) if (i)* $Q(0) = 0$*; (ii)* $Q(1) = 1$*; and (iii)* $x > y$ *implies* $Q(x) \geq Q(y)$*.*

Definition 27. *Let Q be a regular nondecreasing fuzzy quantifier, and let* $\mathbf{p}$ *be a weighting vector of dimension N; then, a mapping WOWA: $\mathbb{R}^N \rightarrow \mathbb{R}$ is a* Weighted Ordered Weighted Averaging (WOWA) operator *of dimension N if*

$$WOWA_{\mathbf{p},Q}(a_1, ..., a_N) = \sum_{i=1}^{N} \omega_i a_{\sigma(i)},$$

where σ is defined as in the case of the OWA, and the weight ω_i is defined as

$$\omega_i = Q\left(\sum_{j \leq i} p_{\sigma(i)}\right) - Q\left(\sum_{j < i} p_{\sigma(i)}\right).$$

There have been implementations of the WOWA operator based on different interpolation methods. In order to make clear the difference between the WOWA operator and the interpolation method, we give a definition in which the interpolation function is a parameter of the operator. We review this definition below.

Definition 28. *[84] Let w be a weighting vector of dimension N with weights $w = (w_1 \ldots w_N)$, let I denote a particular interpolation method, and let $I(w)$ be the function that interpolates the points $\{(i/N, \sum_{j \leq i} w_j)\}_{i=1,\ldots,N} \cup \{(0,0)\}$. Then, we say that I is an interpolation method WOWA-consistent if $I(w)$ is a monotonic function, and $I(w)(x) = x$ when $x = i/N$ for $i = 0, \ldots, N$.*

Naturally, from this definition it follows $I(w)(i/N) = \sum_{j \leq i} w_i$ and $I(w)(0) = 0$. The non-linear interpolation described in [78,8,79] and the linear interpolation (see e.g. [10]) lead to WOWA-consistent interpolation methods. A comparison of the sensitivity of the different interpolation methods is given in [84,81].

Note that taking into account Definition 28, we can define the WOWA operator as follows.

Definition 29. *[84] Let* $\mathbf{p}$ *and* $\mathbf{w}$ *be two weighting vectors of dimension N, let I be a WOWA-consistent interpolation method; then, the mapping WOWA: $\mathbb{R}^N \rightarrow \mathbb{R}$ is a* Weighted Ordered Weighted Averaging (WOWA) operator *of dimension N if*

$$WOWA_{\mathbf{p},\mathbf{w},I}(a_1, ..., a_N) = WOWA_{\mathbf{p},I(w)}(a_1, ..., a_N),$$

Here we have reviewed the definition of the OWA and the WOWA in a discrete domain. For continuous domains see [83,50,51] and Definition 32.

The WOWA operator generalizes the weighted mean and the OWA and, as we will see in the next section, the aggregation operators reviewed in this section can be seen as particular cases of some integrals.

2.4 Choquet and Sugeno Integrals

Let us review the definition of the two main integrals for non-additive measures. We begin with the Choquet integral, that reduces to the Lebesgue integral when the measure is additive. [47] (Chapter 3 in this book) presents an overview of these integrals and other integrals for non-additive measures. [49] (Chapter 2 in this book) also considers these integrals and discuss the case of integration of functions that take negative values. The asymmetric and symmetric (Šipoš) integrals are defined.

Definition 30. *[9] Let X be a reference set, let $(X, \mathcal{A})$ be a measurable space, let μ be a non-additive measure on $(X, \mathcal{A})$, and let f be a measurable function $f : X \to [0, 1]$; then, the Choquet integral of f with respect to μ is defined by*

$$C_\mu(f) := \int_0^\infty \mu_f(r)dr,$$

where $\mu_f(r) := \mu(\{x|f(x) > r\})$.

In the case of a discrete domain X, we have the following equivalent definition for the Choquet integral.

Definition 31. *Let μ be a non-additive measure on X; then, the* Choquet integral *of a function $f : X \to \mathbb{R}^+$ with respect to the fuzzy measure μ is defined by*

$$(C) \int f d\mu = \sum_{i=1}^{N} [f(x_{s(i)}) - f(x_{s(i-1)})]\mu(A_{s(i)}), \tag{4}$$

where $f(x_{s(i)})$ indicates that the indices have been permuted so that $0 \leq f(x_{s(1)}) \leq \cdots \leq f(x_{s(N)}) \leq 1$, and where $f(x_{s(0)}) = 0$ and $A_{s(i)} = \{x_{s(i)}, \ldots, x_{s(N)}\}$.

We can use this integral as an aggregation operator when we consider that f is the function that associates each information source with its value. That is, as above, $f(x_i) = a_i$. In this case, we can use $CI_\mu(a_1, \ldots, a_N)$ to denote $(C) \int f d\mu$.

The WOWA operator is a particular case of the discrete Choquet integral. This relationship was proven in [76]. In particular, for a finite set X the WOWA operator with vectors p, w, and a WOWA-consistent interpolation method I corresponds to a Choquet integral with respect to the distorted probability defined by p and $I(w)$. Equivalently, it corresponds to the Choquet integral with respect to the measure $\mu(A) = I(w)(\sum_{x \in A} p(x))$.

This fact permits us to define the continuous WOWA (CWOWA) using a probability measure P with density function p and a regular nondecreasing fuzzy quantifier Q as follows:

Definition 32. *Let P be a probability measure on $(\mathbb{R}, \mathcal{B})$ with a density function p, that is,*

$$P([a, b]) = \int_{[a,b]} p(x)d\lambda$$

where λ is the Lebesgue measure, let Q be a monotone increasing function on $[0, 1]$, and $f : \mathbb{R} \to \mathbb{R}$ a random variable. Then. Then, the continuous WOWA operator of f is defined by

$$CWOWA_\mu(f) = (C) \int f d\mu$$

where $\mu = Q \circ P$.

Then, the continuous OWA ($COWA$) is defined in terms of a symmetric fuzzy measure. That is, $COWA_\mu(f) = (C)\int f d\mu$ when μ is a symmetric fuzzy measure.

As a consequence of all this, the weighted mean is a particular case of the discrete Choquet integral. In fact, for a finite X, the weighted mean with weights w_i is the Choquet integral with measure $\mu(A) = \sum_{x_i \in A} w_i$. In general, as stated above, the Choquet integral reduces to the Lebesgue integral for additive measures, which corresponds to the expectation of the function being integrated. Similar results are obtained on non finite sets.

Let us now consider the Sugeno integral.

Definition 33. *[68,69] Let X be a reference set, let $(X, \mathcal{A})$ be a measurable space, let μ be a non-additive measure on $(X, \mathcal{A})$, and let f be a measurable function $f : X \to [0,1]$; then, the Sugeno integral of f with respect to μ is defined by*

$$S_\mu(f) := \sup_{r \in [0,1]} [r \wedge \mu_f(r)],$$

where $\mu_f(r) := \mu(\{x | f(x) > r\})$.

The definition for the discrete case follows.

Definition 34. *Let μ be a non-additive measure on X; then, the* Sugeno integral *of a function $f : X \to [0,1]$ with respect to μ is defined by*

$$(S) \int f d\mu = \max_{i=1,N} \min(f(x_{s(i)}), \mu(A_{s(i)})), \tag{5}$$

where $f(x_{s(i)})$ indicates that the indices have been permuted so that $0 \leq f(x_{s(1)}) \leq ... \leq f(x_{s(N)}) \leq 1$ and $A_{s(i)} = \{x_{s(i)}, ..., x_{s(N)}\}$.

Sugeno integrals, due to the combination of weights and values through the minimum, needs the function and the measure to be defined in the same domain. It is usual to consider functions into $[0,1]$ and normalized fuzzy measures.

The Sugeno integral generalizes the weighted minimum and the weighted maximum. See [86] for the definition of these other operators, the relationship between all of them, and some examples of their application.

When the Sugeno integral is used as an aggregation operator, we usually use $SI_\mu(a_1, \ldots, a_N)$ to denote $(S)\int f d\mu$. Here, $a_i = f(x_i)$ as before.

Some Generalizations. We review in this section two level-dependent non-additive integrals, and the twofold integral. The former is defined with respect to a level-dependent fuzzy measure (see Definition 19). In the definitions $\vee$ is the maximum (or supremum) and $\wedge$ the minimum.

Definition 35. *[25] Let $X = (x_1, \ldots, x_N)$ be the reference set, $f : X \to (\alpha, \beta) \subseteq [0,1]$, and let μ be a level dependent capacity. Then, the level dependent Choquet integral of f with respect to μ is defined by:*

$$CI_\mu(f) = \int_0^\infty \mu(A_f(t), t) dt, \tag{6}$$

where

$$A_f(t) = \{x \in X : f(x) \geq t\}.$$

Definition 36. *[46] Let $X = (x_1, \ldots, x_N)$ be the reference set, $f : X \to (\alpha, \beta) \subseteq [0,1]$, and let μ be a level dependent capacity. Then, the level dependent Sugeno integral of f with respect to μ is defined by:*

$$SI_\mu(f) = \bigvee_{t \in (0,1]} \mu(\{x \in X : f(x) \geq t\}, t)$$

Let us consider now the twofold integral.

Definition 37. *[54,80] Let X be a reference set, let $(X, \mathcal{A})$ be a measurable space, and let μ_C and μ_S be non-additive measures on $(X, \mathcal{A})$. Then, for a measurable function $f : X \to [0,1]$, let us define $\phi_f : [0,1] \to [0,1]$ by*

$$\phi_f(x) := \bigvee_{0 \leq r \leq x} (r \wedge \mu_S(\{f > r\})).$$

Note that $\phi_f(1) = S_{\mu_S}(f)$ and ϕ_f is nondecreasing, so the cardinality of noncontinuous points of ϕ_f is at most countable. ϕ_f permits us to define a Lebesgue-Stjeltjes measure ν_{ϕ_f} on the real line by

$$\nu_{\phi_f}([a,b]) := \phi_f(b+0) - \phi_f(a-0).$$

Then, the twofold integral *of a measurable function $f : X \to [0,1]$ with respect to fuzzy measures μ_S and μ_C is defined by*

$$TI_{\mu_S, \mu_C}(f) = \int_0^1 \mu_C(f > a) d\nu_{\phi_f}(a).$$

For other integrals and generalizations of the integrals here see [49] and [47] (Chapters 2 and 3 of this book).

2.5 Inequalities and Bounds

The literature on non-additive measures and integrals has also studied the conditions that are required so that some well known equations and inequalities for additive measures hold for non-additive ones. See e.g. [61,62] for some results in the area and [56] for an overview of the topic.

2.6 Uncertainty Measures

The literature presents a considerable number of approaches with the generic objective of capturing the uncertainty of a probability or, in general, of a non-additive measure. They are the uncertainty measures. A well known example is entropy. We will review a few of them in this section. See [32] (Chapter 4 in this book) for a detailed discussion of the entropy for non-additive measures.

Following [36,37][1], the measures of uncertainty can be classified in three classes.

On the one hand we have the measures that focus on the ambiguity we have in a piece of information. Ambiguity can be due to imprecision (or nonspecificity) or due to discord (or conflict or strife). We have imprecision when there are different possible alternatives. In this case, nonspecificity is to measure the size of the set of possible alternatives. In contrast, other measures as entropy and strife evaluate the degree of conflict, inconsistency or disagreement between these alternatives. According to [27], "the more even the strength of the disagreeing pieces of evidence is, the larger the conflict". The term randomness is also used for this latter set of measures (see e.g. Yager [93]). As we see below, nonspecificity is zero for probabilities, and only entropy and other measures of conflict are able to capture the uncertainty due to randomness.

On the other hand we have uncertainty due to fuzziness or vagueness because of imprecise boundaries of sets ("lack of sharpness of relevant distinctions" as [37] puts it (p. 268)). One approach to define measures of fuzziness is to take into account that the larger the distinction between a fuzzy set and its complement, the smaller its fuzziness. See [36,37] for details on measures of fuzziness and [34] for references.

In this section we review the measures for imprecision (nonspecificity) and conflict (entropy and strife). Measures of fuzziness are not discussed here. The literature presents other measures that aggregate nonspecificity and conflict. They are known as aggregate uncertainty measures (see for details [27,36]).

Nonspecificity and Imprecision. One of the first measures of information was the one defined by Hartley in 1928 [29]. This measure, based on (crisp) set theory, is about the selection of an element from a set of candidates.

Let us consider a set A (subset of X) and a selection of m elements from this set. Then, there are $|A|^m$ sequences of m elements. Then, the basic idea of Hartley's definition is that the uncertainty of m selections should be proportional to m. That is,

$$H(|A|^m) = m \cdot K(|A|), \tag{7}$$

where $K(|A|)$ is a constant that depends on $|A|$ (the cardinality of the set A).

Let us now consider two sets A_1 and A_2, and two values m_1 and m_2 such that $|A_1|^{m_1} = |A_2|^{m_2}$. From this last equation, Hartley derives:

$$m_1 \log |A_1| = m_2 \log |A_2|. \tag{8}$$

In addition, using Equation 7 we have that $H(|A_1|^{m_1}) = H(|A_2|^{m_2})$ corresponds to

$$m_1 \cdot K(|A_1|) = m_2 \cdot K(|A_2|). \tag{9}$$

[1] In [36], information is "conceived in terms of uncertainty reduction".

Equations 8 and 9 imply that $K(|A|) = K_0 \log |A|$. Since K_0 is arbitrary, Hartley removes it making the logarithmic base arbitrary. That is,

$$K(|A|) = \log_b |A|, \tag{10}$$

and $H(|A|^m) = m \log_b |A|$.

When $b = 2$, the uncertainty is measured in bits.

In 1970, Rényi [60] characterized Hartley's measure in terms of three axioms: additivity ($H(n \cdot m) = H(n) \cdot H(m)$), monotonicity ($H(n) \leq H(n+1)$), and normalization ($H(2) = 1$). That is, the only measure that satisfies additivity, monotonicity, and normalization is Equation 10.

In Equation 10, A is the set of alternatives, and the uncertainty is related to the imprecision of the set. Klir and Wierman [36], Dubois and Prade [12] and others use the term nonspecificity for this type of measure. Naturally, full specificity (minimum imprecision) is achieved when we have only one alternative ($|A| = 1$). Maximum nonspecificity (maximum imprecision) is achieved when $A = X$.

In the area of evidence theory, the following definition was introduced by Dubois and Prade [12] to compute the nonspecificity of an arbitrary non-additive measure. A characterization of this measure was given by Ramer in [59].

Definition 38. *Let μ be a non-additive measure and let m be the Möbius transform of μ (see Definition 9), then we define the nonspecificity of μ by:*

$$N(m) = \sum_{A \in \mathcal{F}_\mu} m(A) \log_2 |A| \tag{11}$$

where $\mathcal{F}_\mu$ is the set of focal elements of μ.

When μ is a probability, the nonspecificity of μ is zero. The maximum of this function is achieved for a measure μ with Möbius transform $m(X) = 1$ and $m(A) = 0$ for all other $A \neq X$.

Abellán and Moral [1] generalize Definition 38 to the case of convex sets of probability distributions[2]. Given a convex set of probability distributions $\mathcal{C}$ they define the function $f(A) = \inf_{p \in \mathcal{C}} P(A)$ (this corresponds to the lower non-additive measure of $\mathcal{C}$) and then define the nonspecificity of $\mathcal{C}$ as the nonspecificity of the Möbius transform of f.

The measures discussed in this section are only applicable to finite sets X. The case of nonspecificity when the sets are on the real line or, in general, in the n-dimensional Euclidean space was explored by Klir and Yuan [38].

Their definition is based on a Hartley-like function HL that can be applied to any convex subset A of a set $X \subseteq \mathbb{R}^n$. That is, let $\mathcal{C}$ be the family of all convex subsets of X, then a Hartley-like function has the following form:

$$HL : \mathcal{C} \to \mathbb{R}^+.$$

[2] Recall that given a belief function Bel on X, the closed convex set of probability distributions $\mathcal{C}_{Bel}$ is defined by all probabilities $\{(p_x)|x \in X\}$ satisfying (i) $p_x \in [0,1]$ for $x \in X$ and $\sum p_x = 1$; and (ii) $Bel(A) \leq \sum_{x \in A} p_x \leq 1 - Bel(\bar{A})$ for all $A \subseteq X$ and where $\bar{A}$ is the complement of A.

An example of such function is the following one:

$$HL(A) = \min_{t \in T} \ln \left[\prod_{i=1}^{n} [1 + \lambda(A_{i_t})] + \lambda(A) - \prod_{i=1}^{n} [\lambda(A_{i_t})] \right]$$

where λ denotes the Lebesgue measure, T denotes the set of all transformations from one orthogonal coordinate system to another, and A_{i_t} denotes the ith projection of A within the coordinate system t. This function satisfies the set of properties that Hartley-like functions are required to satisfy in [38,36] (see these references for details). One of the properties is that the function HL has to be coordinate invariant. That is, the function does not change under isometric transformations of the coordinate space. That is why in the definition the set T is considered.

Using HL, nonspecificity for non-additive measures on $\mathbb{R}^n$ is defined as follows, when the set of focal elements of the fuzzy measure μ, i.e., $\mathcal{F}_\mu$, is finite.

$$NL(m) = \sum_{A \subseteq \mathcal{F}_\mu} m(A) HL(A) \tag{12}$$

Conflict. The most well-known measure of uncertainty is the Shannon entropy, introduced by Shannon in 1948 [65]. Its definition follows.

Definition 39. *Let μ be a probability measure on a finite reference set X, and m be its probability density function (its Möbius transform) then the Shannon entropy is defined by*

$$H(\mu) = - \sum_{x \in X} m(\{x\}) \log_2 m(\{x\}).$$

with $0 \log 0$ defined as 0 (to allow for probabilities equal to zero).

There are different characterizations of this function using different sets of properties. See e.g. [3,13] for details.

This measure can be seen as a measure of conflict. This is illustrated in [36] with the following expression that is equivalent to the one above for the entropy:

$$H(\mu) = \sum_{x \in X} m(\{x\}) C(x) \tag{13}$$

where $C(x) = -\log_2 [1 - Con(x)]$ and $Con(x) = \sum_{y \neq x} m(\{y\})$.

In this expression, $Con(x)$ corresponds to the probability that conflicts with x (i.e., the one assigned to elements $y \in X$ such that $x \neq y$), and $C(x)$ is a function that is monotonic increasing with respect to $Con(x)$. So, $H(\mu)$ is the average of $Con(x)$ for all x in X, or the expectation of conflict.

In the case of non-additive measures, there are several alternative definitions for measures on finite sets. We review two of them below. They are the lower and upper entropies. [32] (Chapter 4 in this book) reviews these and other measures.

Definition 40. *[43,44] Let μ be a non-additive measure on a finite set $X = \{x_1, \ldots, x_N\}$; then, the lower entropy H_l of μ is defined by*

$$H_l(\mu) := \sum_{i=1}^{N} \sum_{T \subseteq X \setminus \{x_i\}} \gamma_{|T|}(N) h[\mu(T \cup \{x_i\}) - \mu(T)], \tag{14}$$

where h is defined as follows

$$h(x) := \begin{cases} -x \ln x & \text{if } x > 0 \\ 0 & \text{if } x = 0, \end{cases} \tag{15}$$

and

$$\gamma_t(n) := \frac{(n-t-1)!t!}{n!}. \tag{16}$$

Definition 41. *[94] Let μ be a non-additive measure on a finite set $X = \{x_1, \ldots, x_N\}$; then, the upper entropy H_u of μ is defined by*

$$H_u(\mu) := \sum_{i=1}^{N} h\left(\sum_{T \subseteq X \setminus \{x_i\}} \gamma_{|T|}(N)[\mu(T \cup \{x_i\}) - \mu(T)] \right), \tag{17}$$

with $\gamma_t(n)$ defined as above.

In this second definition, the entropy of a fuzzy measure μ corresponds to the entropy of the Shapley value of the measure μ (see Definition 7 in [18], Chapter 6 in this book). The two definitions above satisfy

1. $H_l(\mu) \leq H_u(\mu)$ for all μ;
2. $H_l(\mu) = H_u(\mu)$ if and only if μ is additive.

The comparison of the two entropies is given in [45]. For some other properties of these entropies see [32] (Chapter 4 in this book) and also [86].

Marichal and Roubens introduced in [45] a definition of entropy for non-additive measures $\mu : 2^X \to L$ when X is finite and L is an ordinal scale. Let $L = \{l_0, \ldots, l_r\}$, with $l_0 <_L l_1 <_L \cdots <_L l_r$; then, the ordinal entropy H_L of μ is defined by

$$E_L(\mu) = l_{|R|} - 2,$$

where $R = \{\mu(A) | A \subset X\}$.

Equation 13 suggested other generalizations of entropy. We review the one introduced by Klir and Parviz in [35]. First, note that $Con(x)$ in Equation 13 can be written as

$$Con(x) = \sum_{A \cap \{x\} = \emptyset} m(A).$$

Recall that in this expression m is the Möbius transform of μ (i.e., the probability density function of μ). Then, the strife ST of a non-additive measure μ is defined by:

$$ST(\mu) = -\sum_{A \in \mathcal{F}_\mu} m(A) \log_2 \left(1 - \sum_{B \in \mathcal{F}_\mu} m(B) \frac{|A - B|}{|A|} \right).$$

Definition 39 is on finite sets, the analogous definition not restricted to finite sets is the Boltzmann entropy. Its definition is

$$B(\mu) = -\int_a^b m(x) \log_2 m(x),$$

where m is a probability density function of μ on $[a, b]$. Nevertheless this definition is not an extension of the Shannon entropy. This is so because when we consider a sequence of discrete distributions on $[a, b]$ with an increasing number of terms, the Shannon entropy does not converge to the Boltzmann entropy.

To solve this problem, the Shannon cross-entropy was defined. The definitions follow. They are defined in terms of two probability density functions p and q. First, we give the one for continuous functions and later the one for discrete ones:

$$B(p, q) = \int_a^b p(x) \log_2 \frac{p(x)}{q(x)} dx$$

and

$$H(p, q) = \sum_{x \in X} p(x) \log_2 \frac{p(x)}{q(x)}.$$

Up to our knowledge, entropies for non-additive measures not restricted to finite sets have not been studied in the literature.

3 Decision Making

In this section we review first some models for decision making, and then focus on problems that cannot be solved with the classical models but that can be solved with the models based on non-additive measures. In [55] (Chapter 5 in this book) some representation theorems are given for the models discussed briefly here.

3.1 Classical Expected Utility

In classical expected utility theory we consider a (finite) state space, which we denote by S, and a set X that corresponds to a (finite) set of outcomes. Then, let $\mathcal{A}$ denote an algebra on X. Then $(X, \mathcal{A})$ is a measurable space. Let P be a probability measure on this space, and, thus, $(X, \mathcal{A}, P)$ is a probabilistic space. Let $u : X \to \mathbb{R}^+$ be a utility function.

We define an act as a function from S to X. Let f be one act, then f is a random variable. $\mathcal{F}$ corresponds to the set of acts. That is,

$$\mathcal{F} = \{f | f : S \to X\}.$$

Then, we also have preferences on $\mathcal{F}$, which are denoted by $\prec$. $\prec$ is a weak order on F and some properties are expected on $\prec$. See [55] (Chapter 5 in this book) for details.

Then, in classical expected utility theory, the preferences $\prec$ are represented by a probability on S. In other words, given the tuple $(S, X, \mathcal{F}, \prec)$ we expect to have a probability P on S such that $E(u(f)) < E(u(g))$ when $f \prec g$. Here E is the expectation. Because of that, to represent preferences by expected utility we need to find a (subjective) probability P such that $E(u(f)) < E(u(g))$ for all f and g in $\mathcal{F}$ such that $f \prec g$. Here, naturally,

$$E(u(f)) = \sum_{s \in S} u(f(s))P(\{s\}) = \sum_{x \in X} u(s)P(f^{-1}(x)).$$

Alternatively, if there is a probability P and a utility function u such that

$$E(u(f)) < E(u(g)) \text{ if and only if } f \prec g$$

we say that P and u represent $\prec$.

3.2 Choquet Expected Utility Model

Schmeidler defined in [64] the Choquet expected utility model for decision under uncertainty. This model uses a Choquet integral and a non-additive measure. This model is as follows.

Definition 42. *A decision maker ranks acts in $\mathcal{F}$ according to the Choquet expected utility model if there is a utility function $u : \mathbb{R}^+ \to \mathbb{R}^+$, and a non-additive measure μ such that $f \prec g$ for $f, g \in \mathcal{F}$ if and only if*

$$CI_{u,\mu}(f) < CI_{u,\mu}(g)$$

where

$$C_{u,\mu}(f) = (C) \int u(f) d\mu$$

is the Choquet integral of $u(f)$ with respect to μ.

3.3 Rank-Dependent Expected Utility Model

Quiggin defined in [58] an alternative model based on distorted probabilities. This model also uses a Choquet integral with a non-additive measure. However, in this case, the measure belongs to the family of distorted probabilities (see Definition 14).

Definition 43. *A decision maker ranks acts in $\mathcal{F}$ according to the rank-dependent expected utility model if there is a utility function $u : \mathbb{R}^+ \to \mathbb{R}^+$, a probability distorting function w (i.e., a nondecreasing function $w : [0,1] \to [0,1]$), and a non-additive probability P such that $f \prec g$ for $f, g \in \mathcal{F}$ if and only if*

$$J_{u,w,P}(f) < J_{u,w,P}(g)$$

where

$$J_{u,w,P}(f) = (C) \int u(f) d(w \circ P).$$

3.4 Ellsberg Paradox

In this section we define Ellsberg paradox and we show how non-additive measures can be used to solve this problem. This problem was defined in [14], a solution using the Choquet integral can be found also in [52].

Example 2. Let us consider an urn with balls of three colours: red, black, and yellow. The number of red balls is 30. The exact number of black and yellow balls is not known, but their total is 60. Therefore, $S = \{R, B, Y\}$.

Let f_R mean that you will get \$ 100 only if you take a red ball and f_B mean that you will get \$ 100 only if you take a black ball. Under these alternatives, most people prefer f_R. Therefore, we have $f_B \prec f_R$. Now, let f_{RY} mean that you will get \$ 100 if you take either a red or a yellow ball, and let f_{BY} mean that you will get \$ 100 if you take a black or a yellow ball. Under these alternatives, most people select f_{BY}. Therefore, $f_{RY} \prec f_{BY}$.

To complete the formalization, we have $X = \{0, 100\}$. Table 1 represents the acts f_B, f_R, f_{RY}, f_{BY}.

Table 1. Balls and acts in Ellsberg's paradox

Color of balls	Red	Black	Yellow
Number of balls	30	60	
f_R	\$ 100	0	0
f_B	\$ 0	\$ 100	0
f_{RY}	\$ 100	0	\$ 100
f_{BY}	\$ 0	\$ 100	\$ 100

This problem cannot be formalized by means of expected utility theory. Note that in order to get a representation of $\prec$ in terms of a utility function and a probability distribution, we need that

$$E(u(f)) \leq E(u(g)) \text{ for all } f \prec g.$$

Therefore, from $f_{RY} \prec f_{BY}$ we have that

$$\begin{aligned} E(u(f_{RY})) &= u(0)P(B) + u(100)P(Y) + u(100)P(R) \\ &< u(100)P(B) + u(100)P(Y) + u(0)P(R) = E(u(f_{BY})) \end{aligned}$$

or, equivalently,

$$u(0)P(B) + u(100)P(R) < u(100)P(B) + u(0)P(R) \tag{18}$$

On the other hand, from $f_B \prec f_R$ we have that

$$\begin{aligned} E(u(f_B)) &= u(100)P(B) + u(0)P(Y) + u(0)P(R) \\ &< u(0)P(B) + u(0)P(Y) + u(100)P(R) = E(u(f_R)) \end{aligned}$$

or, equivalently,

$$u(100)P(B) + u(0)P(R) < u(0)P(B) + u(100)P(R). \tag{19}$$

So, as inequality 19 is in contradiction with inequality 18, it is not possible to find functions u and P according to $\prec$.

In contrast, the Choquet integral utility model can be used to solve this paradox. In particular, we can use the Choquet integral with the following measure to represent the preferences.

- $\mu(\emptyset) = 0$
- $\mu(\{R\}) = 1/3$, $\mu(\{B\}) = \mu(\{Y\}) = 2/9$
- $\mu(\{R, Y\}) = 5/9$, $\mu(\{B, Y\}) = \mu(\{R, B\}) = 2/3$
- $\mu(\{R, B, Y\}) = 1$

The following computations illustrate that the paradox is solved because it is now possible to find u such that for all $f \prec g$ holds

$$CI_\mu(u(f)) < CI_\mu(u(g))$$

This is the case, for example, for $u(x) = x$. In particular, from $f_{RY} \prec f_{BY}$ we have

$$\begin{aligned} CI_\mu(u(f_{RY})) &= u(0)\mu(\{B\}) + u(100)\mu(\{Y, R\}) \\ &< u(100)\mu(\{B, Y\}) + u(0)\mu(\{R\}) = CI_\mu(u(f_{BY})) \end{aligned}$$

or, equivalently,

$$0 \cdot 2/9 + 100 \cdot 5/9 < 100 \cdot 2/3 + 0 \cdot 1/3.$$

On the other hand, from $f_B \prec f_R$ we have that

$$\begin{aligned} CI_\mu(u(f_B)) &= u(100)\mu(\{B\}) + u(0)\mu(\{Y, R\}) \\ &< CI_\mu(u(f_R)) = u(0)\mu(\{B, Y\}) + u(100)\mu(\{R\}) \end{aligned}$$

or, equivalently,

$$100 \cdot 2/9 + 0 \cdot 5/9 < 0 \cdot 2/3 + 100 \cdot 1/3.$$

In [55] (Chapter 5 in this book), it is discussed, following [15], that a submodular non-additive measure is not necessary nor sufficient for explaining Ellsberg's paradox. Examples are given to illustrate these facts. Ellsberg paradox is an example of ambiguity aversion[3].

---

[3] Ambiguity aversion or uncertainty aversion describes an attitude of preference for known risks over unknown risks.

4 Subjective Evaluations

Another area where integrals have been applied is the area of subjective evaluation. This topic already appears as one of the motivations in Sugeno's work [69] (p. 2).

> "The purposes of this dissertation are to propose the concept of fuzzy measures and integrals [11,12] as a way for expressing human subjectivity and to discuss their applications."

Dubois and Prade [11] define subjective evaluation as follows.

> "Formally speaking, the subjective evaluation problem can be viewed as the synthesis, the identification of a function which maps the attribute values describing the situation to evaluate into a discrete domain (classification), or a continuous one (absolute evaluation). More generally, we may look for the degree of membership of the situation to a category, or have a function yielding a fuzzy evaluation. This function is in general not available as such, but is implicitly, and partially, described in terms of criteria, or by means of expert rules, or through some fuzzy algorithm. It may also happen that the function is only partially known by exemplification through prototypical examples of situations for which the evaluation is available."

One of the approaches to subjective evaluation is to consider multiple criteria, and aggregate them by means of an aggregation function. Using the notation of Section 2.3, we have a set of finite criteria $X = \{x_1, \ldots, x_N\}$, a set of functions, one for each object to be evaluated f_i, and then the aggregation function $\mathbb{C}$ to combine the values $f_i(x_1), \ldots, f_i(x_N)$.

As described in Section 2.3, several functions exist for combining these values. The weighted mean is one of them. In this case, the weights correspond to the importance of the criteria (i.e., p_i corresponds to the importance of criteria x_i). Then, an object with evaluation f_i is prefered to another object f_j with respect to weights $p_1, \ldots, p_N$ when $WM_p(f_i(x_1), \ldots, f_i(x_N)) \geq WM_w(f_j(x_1), \ldots, f_j(x_N))$. As stated in Section 2.3, the weighted mean corresponds to the Lebesgue integral. That is, an integral with an additive measure.

Several examples have been presented in the literature to show the limitations of models based on additive measures to represent preferences. We review some of them below. We describe an example introduced by Grabisch [21] and another by Greco et al. [25]. The first one can be solved with the Choquet integral, but not with a weighted mean. In the second case, the Choquet integral is not suitable but, in contrast, the level-dependent Choquet integral can be used.

4.1 Grabisch's Example

The example consists on the evaluation of three students in a school, according to the marks on three subjects. The goal is to represent a subjective evaluation of the director of the school.

Table 2. Marks of the students in mathematics (M), physics (P), and literature (L) according to Grabisch [21]

Student		M	P	L
Ada	f_A	18	16	10
Byron	f_B	10	12	18
Countess	f_C	14	15	15

Example 3. Let us consider three students A, B, and C (for Ada, Byron, and the Countess) and their marks in three subjects *Mathematics*, *Physics*, and *Literature* according to Table 2 (marks in a scale of $[0, 20]$). The preferences of the director of the school are as follows:

- The director wants to assign the same weight to mathematics and physics, and more weight to this subjects than to literature. This is so because the school is *scientifically* oriented.
- The director wants to represent the following preference on the students:

$$B \prec A \prec C. \tag{20}$$

  This preference is based on the fact stated above that the school is more *scientifically* than literary oriented, and, thus, more importance is attributed to mathematics and physics than to literature. So, in short, the director prefers A and C to B. Nevertheless, as for the director C and A are equally good at scientific subjects, C is prefered to A because C is also good in literature.

If we try to model the ranking of the director in terms of a weighted mean of the marks of the students, we have

$$E(s) = w_M \cdot m_M(s) + w_P \cdot m_P(s) + w_L \cdot m_L(s)$$

where $m_M(s)$ is the mark of the student s in mathematics, $m_P(s)$ is the mark of the student s in physics, and $m_L(s)$ is the mark of the student s in literature, and where w_M, w_P, and w_L are the weights of the subjects (mathematics, physics, and literature). In this case, it is impossible to find weights so that $E(B) < E(A) < E(C)$ and at the same time $w_M = w_P > w_L$[4].

In contrast to the weighted mean, the Choquet integral permits the representation of directors preferences. For this purpose we can use the fuzzy measure defined as follows:

- $\mu(\emptyset) = 0$, $\mu(\{M, P, L\}) = 1$
- $\mu(\{M\}) = \mu(\{P\}) = 0.45$, $\mu(\{L\}) = 0.3$

[4] Note that there are weights that satisfy Equation 20 but then the other conditions are not satisfied. This is the case e.g. of the weights $w_M = 1/4$, $w_P = 1.5/4$ and $w_L = 1.5/4$ which imply $E(B) = 13.75 < E(A) = 14.25 < E(C) = 14.75$ but then $w_M < w_L$.

- $\mu(\{M,P\}) = 0.5 < \mu(\{M\}) + \mu(\{P\})$
- $\mu(\{M,L\}) = \mu(\{P,L\}) = 0.9 > 0.45 + 0.3$

Then, we define the evaluation of a student using the Choquet integral of the corresponding function with respect to this fuzzy measure. That is, $E(s) = (C)\int f_s d\mu$. In our example, this leads to

- $E(Ada) = (C)\int f_A d\mu = 13.9$
- $E(Byron) = (C)\int f_B d\mu = 13.6$
- $E(Countess) = (C)\int f_C d\mu = 14.9$

which satisfies the preferences expressed in Equation 20.

4.2 Greco, Matarazzo and Giove's Example

A kind of extension of the example presented by Grabisch [21] (Example 3 above) was introduced by Greco et al. in [25]. In this case, the director has a set of preferences on 8 students.

Example 4. Let $S = \{s_1, s_2, s_3, s_4, s_5, s_6, s_7, s_8\}$ be a set of students with marks on the subjects mathematics, physics, and literature according to Table 3.

Table 3. Marks of the students in mathematics (M), physics (P), and literature (L) according to Greco, Matarazzo, and Giove [25]

Student	M	P	L
S_1	28	28	27
S_2	27	28	28
S_3	26	26	25
S_4	25	26	26
S_5	23	23	22
S_6	22	23	23
S_7	19	19	18
S_8	18	19	19

For these marks, the subjective evaluation of the director of the school is as follows:

$$S_7 \prec S_8 \prec S_6 \prec S_5 \prec S_3 \prec S_4 \prec S_2 \prec S_1$$

For this problem, there is no set of weights that permits to represent director's preference in terms of an average (weighted mean) of the marks with respect to the weights. Note that $E(S_2) \prec E(S_1)$ implies $w_M > w_L$, but $E(S_3) \prec E(S_4)$ implies $w_M < w_L$. In fact, the Choquet integral is neither appropriate to represent this problem. No fuzzy measure exists to express this preference using a Choquet integral. This is so because $CI(s_1) > CI(s_2)$ implies

$$\begin{aligned} CI(s_1) &= 27 \cdot \mu(\{M,P,L\}) + 1 \cdot \mu(\{M,P\}) \\ &> 27 \cdot \mu(\{M,P,L\}) + 1 \cdot \mu(\{P,L\}) = CI(S2) \end{aligned}$$

This means $\mu(\{M,P\}) > \cdot\mu(\{P,L\})$, and, in contrast, $CI(s_3) < CI(s_4)$ implies

$$\begin{aligned} CI(s_3) &= 25 \cdot \mu(\{M,P,L\}) + 1 \cdot \mu(\{M,H\}) \\ &< 25 \cdot \mu(\{M,P,L\}) + 1 \cdot \mu(\{P,L\}) = CI(s_4) \end{aligned}$$

This means $\mu(\{M,P\}) < \mu(\{P,L\})$, which is in contradiction with the previous equality.

Greco et al. prove in [25] that the level-dependent Choquet integral can be used to express the preferences, but that two other alternative models are not adequate. In particular, they prove that neither the bipolar Choquet integral [22,23] nor the Cumulative Prospect Theory functional [91] are valid approaches.

5 Other Applications

There has been a large number of other applications of fuzzy integrals. The fields of computer vision and information fusion have used these integrals for different purposes. We review some of these applications below. Other applications not discussed here include computer vision and fuzzy inference (control). [33] reviews applications in computer vision and [83,63,85] are on applications to different aspects of fuzzy inference and control. In particular, [83] focuses on defuzzification, and [63,85] on the computation of the output fuzzy set.

5.1 Integrals and Data Fusion

[19] considers the problem of land mine detection. The authors describe a system that collects data from multiple sensors and then fuse the information using two soft computing approaches. The output of the system is a mine confidence value. That is, a confidence that a mine is present at a particular location. One approach uses fuzzy logic rules (a Mamdani system [42]), and the other a fuzzy integral. In the former approach two fuzzy rule systems are implemented and running in parallel. The authors state that one is for *linked* or spatially correlated targets and the other for *unlinked* targets. In the other approach, both Choquet and Sugeno integral were considered in the fusion process. Measures were determined heuristically. The authors use Sugeno λ-measures and they interpret the measure of a singelton as the "(possibly subjective) importance of a single information source in determining the evaluation of land mine confidence". Although the fuzzy measures used in this work are determined heuristically, the authors describe that in [20] they compared the use of training algorithms based on quadratic programming versus an heuristic assignment.

The problem of gray level and color image segmentation has been considered by different authors considering non-additive integrals. See e.g. [33]. Most of the papers use Sugeno λ-measures determined from the measures on the singletons. The problem of color image segmentation is also considered in [67]. In this case, the author also considers the use of a Choquet integral with a Sugeno λ-measure. One of the interests of this paper is that they determine the measure by means of an unsupervised approach based on a self-organizing feature map (SOFM) [39]. [40] is another example of unsupervised approach.

Recently, the Choquet integral has been used for regression. See [72,73] for details. This problem can be seen both as an application or as a way to determine the corresponding non-additive measure.

5.2 Identification of Measures

The problem of measure determination (model fitting) briefly mentioned in Section 5.1 is reviewed in more detail in [86]. Basically, we have heuristic methods, and methods that extract the measure from examples. Among the former, we might have an expert that just assigns the values of the measure for all subsets, or an expert that assigns a few relevant parameters from which to extract these values. An example of the later is when the measure is determined as the one that maximises the entropy for a given orness. That is, the expert settles the orness, and the measure is found from this value. Among the later, we have supervised approaches and unsupervised ones. In the supervised case, we have some examples with known input-output pairs, which are used to learn the values of the measure. The case described above using a self-organizing feature map to determine the measure is an example of unsupervised approach.

5.3 Integrals and Distances

Pham and Yan in [57][5] use the Choquet integral to compute the distance between pairs of objects in a clustering algorithm (the mountain clustering algorithm [95]). Their application, on the segmentation of color image data, defines a distance between a point x and a cluster center p by $d(x,p) = 1 - \sigma^*(x,p)$, where the similarity σ^* is computed using the Choquet integral. The authors report that both Choquet and Sugeno integrals were considered and the solutions obtained from the Choquet integral were found better than the ones of the Sugeno integral. Apparently, $\sigma(x,p)$ is computed as the Choquet integral of the three RGB colours of a pixel. That is,

$$\sigma(x,p) = CI_\mu((R_x - R_p)^2, (G_x - G_p)^2, (B_x - B_p)^2). \tag{21}$$

The integration is done with respect to a Sugeno λ-measure.

The use of Choquet integral as a fundamental brick to build a distance funcion has been used in other contexts. In [2] the Choquet integral is used to compare

[5] The authors also discuss in their paper three interpretations for the fuzzy integral: (i) fuzzy expectation (according to [70]), (ii) the maximal degree of agreement between two opposite tendencies (according to [92]), (iii) the maximal grade of agreement between the objective evidence and the expectation (following [71]). Nevertheless, they state that in their paper a fuzzy integral is considered as a maximum degree of belief for an object to belong to a certain class.

a set of numerical records against a given one in order to find the nearest one. Formally, given two records $a = (a_1, \ldots, a_N)$ and $b = (b_1, \ldots, b_N)$ where a_i and b_i are values for variable V_i in $\mathbb{R}$ (and thus, a and b are in $\mathbb{R}^N$), the distance between a and b is defined by

$$d_\mu(a,b)^2 = CI_\mu = ((a_1 - b_1)^2, \ldots, (a_N - b_N)^2) \tag{22}$$

where μ is a fuzzy measure defined on the set of variables $V_1, \ldots, V_N$. Note that this is the same expression given in Equation 21 above.

This function does not satisfy the triangle inequality but only non-negativity, reflexivity, and symmetry. Note also that Equation 22 corresponds to the square of the distance (i.e., $d(a,b)^2 = d(a,b) \cdot d(a,b)$).

Abril et al.[2] explains how to determine (i.e., learn according to machine learning jargon) the measure for the distance in Equation 22 from a set of examples. More specifically, the authors solve a record linkage problem [16]. In short, there are two files A and B both containing the same data from the same individuals in terms of the same variables $V_1, \ldots, V_N$. Records in A and B are not exactly the same due to errors in the data. The problem consists of assigning each record in A to the correct record in B. For a given record a_i, the distance $d(a_i, b)$ is computed for all $b \in B$ and the record with a minimal distance is assigned to a_i. As stated above $d_\mu(a,b)$ is the distance of Equation 22. Given A, B and an assignment of records from A to B, the optimal measure μ is found that corresponds to the assignment that minimizes the number of incorrect links.

Narukawa in [48] also discusses distances defined by a Choquet integral. Let μ be a non-additive measure on $(X, \mathcal{A})$, and let $C_p(\mu)$ be the following subset of measurable functions

$$C_p(\mu) := \left\{ f \in \mathcal{M} | (C) \int |f|^p d\mu < \infty \right\},$$

where $\mathcal{M}$ is the class of measurable functions. Then, [48] introduces the pth power norm of a function f for $p \geq 1$ for submodular μ as

$$||f||_{\mu,p} = \left((C) \int |f|^p d\mu \right)^{1/p}.$$

and prove that if μ is submodular and continuous from below, then the space $(C_p(\mu), ||f||_{\mu,p})$ is complete. This result is later used to define distances on fuzzy sets.

The properties of the distance based on the Choquet integral are also discussed in [7], where the authors characterize the class of measures that induce a metric with the Choquet integral.

A different topic is the definition of distances for non-additive measures. In the case of additive measures we have distances as the Hellinger distance [31], the Kullback and Leibler divergence and the f-divergence. In [89] a definition of the Hellinger distance for non-additive measures was proposed.

5.4 Integrals and Distributions

Expression 22 has been used in [88] in another context to define a probability distribution based on the Choquet integral. More specifically, the exponential family of Choquet integral based class-conditional probability-density functions is defined by:

$$P(x) = \frac{1}{K} e^{-\frac{1}{2} CI_{\mu}((x-m) \otimes (x-m))}$$

where K is a constant so that the function is a probability, and where $v \otimes w$ denotes the elementwise product of vectors v and w (i.e., $(v \otimes w) = (v_1 w_1 \dots v_n w_n)$).

This distribution has similarities with the multinomial normal distribution in the sense that both are multivariate and permit us to represent interactions (and non-independence) between the variables. Nevertheless, the type of interactions they represent are different. Because of that a more general distribution was also introduced, the exponential family of Choquet-Mahalanobis integral based class-conditional probability-density functions is defined by:

$$P(x) = \frac{1}{K} e^{-\frac{1}{2} CI_{\mu,Q}(v \otimes w)}$$

where K is a constant that is defined so that the function is a probability, where $LL^T = Q$ is the Cholesky decomposition of the matrix Q, $v = (x-m)^T L$, $w = L^T(x-m)$, and where $v \otimes w$ denotes the elementwise product of vectors v and w.

6 Conclusions

In this chapter we have reviewed several definitions related to non-additive measures and some of their results. We have briefly discussed some applications. We have shown that non-additive measures and integrals solve some of the short-commings of alternative models.

As a summary, we have seen that non-additive measures when combined with Choquet integrals have more expressive capabilities than additive measures with the Lebesgue integral, or in the discrete setting that the Choquet integral has better modeling capabilities than the weighted mean.

Although no discussed here, it is important to point out that in practical problems, it is not always the best option to select the model with better modeling capabilities because it can cause overfitting (see e.g. [30]). In practice, it is important to find a good trade-off between the simplicity of the model and the complexity of the data or the situation to be represented.

Acknowledgments. Partial support by the Generalitat de Catalunya (2005 SGR 00446 and 2005-SGR-00093) and by the Spanish MEC (projects ARES – CONSOLIDER INGENIO 2010 CSD2007-00004, eAEGIS – TSI2007-65406-C03-02, and co-privacy – TIN2011-27076-C03-03) is acknowledged. The weights for the weighted mean in Example 3 were computed by M. Torra.

References

1. Abellán, J., Moral, S.: A non-specificity measure for convex sets of probability distributions. In: 1st Int. Symp. on Imprecise Probabilities and Their Applications (1999)
2. Abril, D., Navarro-Arribas, G., Torra, V.: Choquet Integral for Record Linkage. Annals of Operations Research 195, 97–110 (2012)
3. Aczél, J., Daróczy, Z.: On Measures of Information and their Characterizations. Academic Press (1975)
4. Beliakov, G.: Shape preserving splines in constructing WOWA operators. Fuzzy Sets and Systems 121(3), 549–550 (2001)
5. Beliakov, G., Pradera, A., Calvo, T.: Aggregation functions: a guide for practitioners. Springer (2007)
6. Bilbao, J.M.: Cooperative games on combinatorial structures. Kluwer Academic Publishers (2000)
7. Bolton, J., Gader, P., Wilson, J.N.: Discrete Choquet Integral as a Distance Metric. IEEE Trans. on Fuzzy Systems 16(4), 1107–1110 (2008)
8. Chen, J.E., Otto, K.N.: Constructing membership functions using interpolation and measurement theory. Fuzzy Sets and Systems 73(3), 313–327 (1995)
9. Choquet, G.: Theory of capacities. Ann. Inst. Fourier 5, 131–295 (1953/1954)
10. Damiani, E., di Vimercati, S.D.C., Samarati, P., Viviani, M.: A WOWA-based aggregation technique on trust values connected to metadata. Electronic Notes in Theoretical Computer Science 157, 131–142 (2006)
11. Dubois, D., Prade, H.: Fuzzy Criteria and Fuzzy Rules in Subjective Evaluation - A General Discussion. In: Proc. 5th European Congress on Intelligent Technologies and Soft Computing (1997)
12. Dubois, D., Prade, H.: A note on measures of specificity for fuzzy sets. Int. J. of General Systems 10(4), 279–283 (1985)
13. Ebanks, B., Sahoo, P., Sander, W.: Characterizations of Information Measures. World Scientific (1997)
14. Ellsberg, D.: Risk, ambiguity, and the savage axioms. Quarterly Journal of Economics 4, 643–669 (1961)
15. Epstein, L.G.: A Definition of Uncertainty Aversion. Review of Economic Studies 66, 579–608 (1999)
16. Fellegi, I.P., Sunter, A.B.: A theory of record linkage. Journal of the American Statistical Association 64, 1183–1210 (1969)
17. Flaminio, T., Godo, L., Kroupa, T.: Belief functions on MV-algebras of fuzzy sets: an overview. In: Torra, V., Narukawa, Y., Sugeno, M. (eds.) Non-additive Measures: Theory and Applications (2013)
18. Fujimoto, K.: Cooperative game as non-additive measure. In: Torra, V., Narukawa, Y., Sugeno, M. (eds.) Non-additive measures: theory and applications (2013)
19. Gader, P.D., Keller, J.M., Nelson, B.N.: Recognition technology for the detection of buried land mines. IEEE Trans. on Fuzzy Systems 9(1), 31–43 (2001)
20. Gader, P.D., Nelson, B., Hocaoglu, A.K., Auephanwiriyakul, S., Khabou, M.: Neural versus heuristic development of choquet fuzzy integral fusion algorithms for land mine detection. In: Bunke, H., Kandel, A. (eds.) Neuro-Fuzzy Pattern Recognition. World Scientific (2000)
21. Grabisch, M.: Fuzzy integral in multicriteria decision making. Fuzzy Sets and Systems 69, 279–298 (1995)

22. Grabisch, M., Labreuche, C.: Bi-capacities - Part I: definition, Möbius transform and interaction. Fuzzy Sets and Systems 151, 211–236 (2005)
23. Grabisch, M., Labreuche, C.: Bi-capacities - Part II: the Choquet integral. Fuzzy Sets and Systems 151, 237–259 (2005)
24. Grabisch, M., Marichal, J.-L., Mesiar, R., Pap, E.: Aggregation Functions, Cambridge University Press. Encyclopedia of Mathematics and its Applications, No. 127 (2009)
25. Grecko, S., Matarazzo, B., Giove, S.: The Choquet integral with respect to a level dependent capacity. Fuzzy Sets and Systems 175, 1–35 (2011)
26. Halmos, P.R.: Measure theory. Springer (1974)
27. Harmanec, D.: Measures of uncertainty and information, Society for Imprecise Probability Theory and Applications (1999), `http://www.sipta.org`
28. Harmanec, D., Klir, G.: Measuring total uncertainty in Dempster-Shafer theory: a novel approach. Int. J. of Intel. Systemss 22, 405–419 (1994)
29. Hartley, R.V.L.: Transmission of information. The Bell Systems Technical Journal 25(2), 153–163 (1928)
30. Hastie, T., Tibshirani, R., Friedman, J.: The Elements of Statistical Learning. Springer, Berlin (2001)
31. Hellinger, E.: Neue Begründung der Theorie quadratischer Formen von unendlichvielen Veränderlichen. Journal für die Reine und Angewandte Mathematik 136, 210–271 (1909)
32. Honda, A.: Entropy of capacity. In: Torra, V., Narukawa, Y., Sugeno, M. (eds.) Non-additive Measures: Theory and Applications (2013)
33. Keller, J.M., Gader, P.D., Hocaoglu, A.K.: Fuzzy Integrals in Image Processing and Recognition. In: Grabisch, M., Murofushi, T., Sugeno, M. (eds.) Fuzzy Measures and Integrals, pp. 435–466. Springer (2000)
34. Klir, G.J., Folger, T.A.: Fuzzy sets, uncertainty, and information. Prentice-Hall (1988)
35. Klir, G.J., Parviz, B.: A note on the measure of discord. In: Dubois, D. (ed.) Proc. of the 8th Conference on Artificial Intelligence, pp. 138–141 (1992)
36. Klir, G.J., Wierman, M.J.: Uncertainty-Based Information: Elements of Generalized Information Theory. Physica-Verlag (1999)
37. Klir, G.J., Yuan, B.: Fuzzy Sets and Fuzzy Logic: Theory and Applications. Prentice Hall, UK (1995)
38. Klir, G.J., Yuan, B.: On nonspecificity of fuzzy sets with continuous membership functions. In: Proc. of the 1995 Int. Conf. on Systems, Man and Cybernetics, Vancouver, pp. 25–29 (1995)
39. Kohonen, T.: Self-Organizing Maps. Springer (1995)
40. Kojadinovic, I.: Unsupervised aggregation by the Choquet integral based on entropy functionals: application to the evaluation of students. In: Torra, V., Narukawa, Y. (eds.) MDAI 2004. LNCS (LNAI), vol. 3131, pp. 163–175. Springer, Heidelberg (2004)
41. Leszczyski, K., Penczek, P., Grochulski, W.: Sugeno's fuzzy measure and fuzzy clustering. Fuzzy Sets and Systems 15, 147–158 (1985)
42. Mamdani, E.H., Assilian, S.: An experiment in linguistic synthesis with a fuzzy logic controller. Int. J. Man-machine Studies 7, 1–13 (1975)
43. Marichal, J.-L.: Aggregation Operators for Multicriteria Decision Aid. Ph. D. Dissertation, Institute of Mathematics, University of Liège, Liège, Belgium (1998)
44. Marichal, J.-L.: Entropy of discrete Choquet capacities. European Journal of Operational Research 137(3), 612–624 (2002)

45. Marichal, J.-L., Roubens, M.: Entropy of discrete fuzzy measures. Int. J. of Unc., Fuzz. and Knowledge Based Systems 8(6), 625–640 (2000)
46. Mesiar, R., Mesiarová-Zemánková, A., Ahmad, K.: Level-dependent Sugeno integral. IEEE Trans. on Fuzzy Systems 17(1), 167–172 (2009)
47. Mesiar, R., Stupňanová, A.: Integral sums and integrals. In: Torra, V., Narukawa, Y., Sugeno, M. (eds.) Non-additive Measures: Theory and Applications (2013)
48. Narukawa, Y.: Distances defined by Choquet integral. In: Proc. of the 2007 IEEE Int. Conf. on Fuzzy Systems (2007)
49. Narukawa, Y.: Integral with respect to a non additive measure: An overview. In: Torra, V., Narukawa, Y., Sugeno, M. (eds.) Non-additive Measures: Theory and Applications (2013)
50. Narukawa, Y.: Torra, Continuous OWA operator and its calculation. In: Proc. IFSA-EUSFLAT, pp. 1132–1135 (2009)
51. Narukawa, Y., Torra, V., Sugeno, M.: Choquet integral with respect to a symmetric fuzzy measure of a function on the real line. Annals of Operations Research (in press, 2013)
52. Narukawa, Y., Murofushi, T.: Choquet Stieltjes integral as a tool for decision modeling. Int. J. of Intel. Syst. 23, 115–127 (2008)
53. Narukawa, Y., Stokes, K., Torra, V.: Fuzzy measures and comonotonicity on multisets. In: Torra, V., Narakawa, Y., Yin, J., Long, J. (eds.) MDAI 2011. LNCS, vol. 6820, pp. 20–30. Springer, Heidelberg (2011)
54. Narukawa, Y., Torra, V.: Twofold integral and Multi-step Choquet integral. Kybernetika 40(1), 39–50 (2004)
55. Ozaki, H.: Integral with respect to Non-additive Measure in Economics. In: Torra, V., Narukawa, Y., Sugeno, M. (eds.) Non-additive Measures: Theory and Applications (2013)
56. Pap, E., Štrboja, M.: Generalizations of integral inequalities for integrals based on nonadditive measures. In: Pap, E. (ed.) Intelligent Systems: Models and Applications. TIEI, vol. 3, pp. 3–22. Springer, Heidelberg (2013)
57. Pham, T.D., Yan, H.: Color image segmentation using fuzzy integral and mountain clustering. Fuzzy Sets and Systems 107, 121–130 (1999)
58. Quiggin, J.: A theory of anticipated utility. J. Econ. Behav. Organ. 3, 323–343 (1982)
59. Ramer, A.: Uniqueness of information measure in the theory of evidence. Fuzzy Sets and Systems 24(2), 185–196 (1987)
60. Rényi, A.: Probability theory. North-Holland (1979)
61. Román-Flores, H., Flores-Franulic, A., Chalco-Cano, Y.: A Jensen type inequality for fuzzy integrals. Information Sciences 177, 3192–3201 (2007)
62. Román-Flores, H., Chalco-Cano, Y.: Sugeno integral and geometric inequalities. Int. J. of Unc., Fuzz. and Knowledge-Based Systems 15(1), 1–11 (2007)
63. Sanz, J., Lopez-Molina, C., Cerrón, J., Mesiar, R., Bustince, H.: A new fuzzy reasoning method based on the use of the Choquet integral. In: Proc. EUSFLAT 2013, pp. 691–698 (2013)
64. Schmeidler, D.: Subjective probability and expected utility without additivity. Econometrica 57, 517–587 (1989)
65. Shannon, C.E.: The mathematical theory of communication. The Bell System Technical Journal 27(3-4) 379–423, 623–656 (1948)
66. Shao, J.: Mathematical Statistics. Springer (2010)
67. Soria-Frisch, A.: Unsupervised construction of fuzzy measures through self-organizing feature maps and its application in color image segmentation. Int. J. of Approx. Reasoning 41, 23–42 (2006)

68. Sugeno, M.: Fuzzy measures and fuzzy integrals. Trans. of the Soc. of Instrument and Control Engineers 8(2) (1972)
69. Sugeno, M.: Theory of Fuzzy Integrals and its Applications, Ph. D. Dissertation, Tokyo Institute of Technology, Tokyo, Japan (1974)
70. Sugeno, M.: Fuzzy measures and fuzzy integrals a survey. In: Gupta, M.M., Saridis, G.N., Gaines, B.R. (eds.) Fuzzy Automata and Decision Processes, pp. 89–102. North-Holland (1977)
71. Tahani, H., Keller, J.M.: Information fusion in computer vision using the fuzzy integral. IEEE Trans. on Systems, Man and Cybernetics 20(3), 733–741 (1990)
72. Tehrani, A.F., Cheng, W., Hüllermeier, E.: Choquistic Regression: Generalizing Logistic Regression using the Choquet Integral. In: Proc. EUSFLAT 2011, pp. 868–875 (2011)
73. Tehrani, A.F., Cheng, W., Dembczyński, K., Hüllermeier, E.: Learning monotone nonlinear models using the Choquet integral. Machine Learning 89, 183–211 (2012)
74. Torra, V.: Weighted OWA operators for synthesis of information. In: Proc. of the 5th IEEE Int. Conf. on Fuzzy Systems, pp. 966–971 (1996)
75. Torra, V.: The weighted OWA operator. Int. J. of Intel. Syst. 12, 153–166 (1997)
76. Torra, V.: On some relationships between the WOWA operator and the Choquet integral. In: Proc. of the IPMU 1998 Conference, pp. 818–824 (1998)
77. Torra, V.: On hierarchically S-decomposable fuzzy measures. Int. J. of Intel. Syst. 14(9), 923–934 (1999)
78. Torra, V.: The WOWA operator and the interpolation function W*: Chen and Otto's interpolation method revisited. Fuzzy Sets and Systems 113(3), 389–396 (2000)
79. Torra, V.: Author's reply to [4]. Fuzzy Sets and Systems 121, 551 (2001)
80. Torra, V.: La integral doble o *twofold* integral: Una generalització de les integrals de Choquet i Sugeno. Butlletí de l'Associació Catalana d'Intel·ligència Artificial 29, 13–19 (2003); Preliminary version in English: Twofold integral: A generalization of Choquet and Sugeno integral, IIIA Technical Report TR-2003-08
81. Torra, V.: Effects of orness and dispersion on WOWA sensitivity. In: Alsinet, T., Puyol-Gruart, J., Torras, C. (eds.) Artificial Intelligence Research and Development, pp. 430–437. IOS Press (2008) (ISBN 978-1-58603-925-7)
82. Torra, V.: The WOWA operator: A review. In: Yager, R.R., Kacprzyk, J., Beliakov, G. (eds.) Recent Developments in the Ordered Weighted Averaging Operators: Theory and Practice. STUDFUZZ, vol. 265, pp. 17–28. Springer, Heidelberg (2011)
83. Torra, V., Godo, L.: Continuous WOWA operators with application to defuzzification. In: Calvo, T., Mayor, G., Mesiar, R. (eds.) Aggregation Operators: New Trends and Applications, pp. 159–176. Physica-Verlag, Springer (2002)
84. Torra, V., Lv, Z.: On the WOWA operator and its interpolation function. Int. J. of Intel. Systems 24, 1039–1056 (2009)
85. Torra, V., Narukawa, Y.: The interpretation of fuzzy integrals and their application to fuzzy systems. Int. J. of Approx. Reasoning 41(1), 43–58 (2006)
86. Torra, V., Narukawa, Y.: Modeling decisions: information fusion and aggregation operators. Springer (2007)
87. Torra, V., Narukawa, Y.: Modelització de decisions: fusió d'informació i operadors d'agregació. UAB Press (2007)
88. Torra, V., Narukawa, Y.: On a comparison between Mahalanobis distance and Choquet integral: The Choquet–Mahalanobis operator. Information Sciences 190, 56–63 (2012)
89. Torra, V., Narukawa, Y., Sugeno, M., Carlson, M.: Hellinger distance for fuzzy measures. In: Proc. EUSFLAT 2013 (2013)

90. Torra, V., Stokes, K., Narukawa, Y.: An Extension of Fuzzy Measures to Multisets and Its Relation to Distorted Probabilities. IEEE Trans. on Fuzzy Systems 20(6), 1032–1045 (2012)
91. Tversky, A., Kahneman, D.: Advances in prospect theory: cumulative representation of uncertainty. Journal of Risk and Uncertainty 5, 297–323 (1992)
92. Wierzchon, S.T.: On fuzzy measure and fuzzy integral. In: Gupta, M.M., Sanchez, E. (eds.) Fuzzy Information and Decision Processes, pp. 79–86. North-Holland (1982)
93. Yager, R.R.: Entropy and specificity in a mathematical theory of evidence. Int. J. of General Systems 9, 249–260 (1983)
94. Yager, R.R.: On the entropy of fuzzy measures, Technical Report #MII-1917R, Machine Intelligence Institute, Iona College, New Rochelle, NY (1999)
95. Yager, R.R., Filev, D.P.: Approximate clustering via the mountain method. IEEE Trans. on Systems, Man and Cybernetics 24, 1279–1284 (1994)

Integral with Respect to a Non Additive Measure: An Overview

Yasuo Narukawa

Toho Gakuen
3-1-10, Naka, Kunitchi, Tokyo, Japan
Department of Computational Intelligence and Systems Science
Tokyo Institute of Technology
4259 Nagatuta, Midori-ku, Yokohama 226, Japan
narukawa@fz.dis.titech.ac.jp

Abstract. This chapter surveys the fundamental aspect of non-additive measures and integral with respect to a non additive measure.

Several basic definitions of non additive measure, Sugeno integral and Choquet integral are presented. The basic properties of the generalized fuzzy integral which is a generalization of both Sugeno and Choquet integral are shown. The generalized Möbius transform and the representation of Choquet integral are also shown.

Keywords: non additive measure, fuzzy measure, Choquet integral, Sugeno integral, Möbius transform.

1 Introduction

A measure is a generalization of the concept of length, area and volume. Suppose $\mu(A)$ denotes the "length" of a set A.

The properties of the notion of "length" are non-negativity and additivity: $\mu(A \cup B) = \mu(A) + \mu(B)$ for $A \cap B = \emptyset$ $A, B \in \mathcal{S}$, where S is a universal set and $\mathcal{S}$ is a class of subsets of S.

In the Lebesgue's measure theory[21], the assumption of additivity is replaced by a countable additivity or $\sigma-$additivity; $\mu(\cup_{i \in I} A_i) = \sum_{i \in I} \mu(A_i)$, where $A_i \in \mathcal{S}, A_i \cap A_j = \emptyset, i, j \in I$,$I$:countable. Lebesgue integral has a lot of useful properties for both theory and application. Especially it is applied for a fundamental background for probability theory [17,19].

However some observation says that probability theory based on Lebesgue measure and integral is too restrictive for Human Centered System. The problem is caused by the additivity of measure and integral. We need another measure theory, that is, non-additive measure theory or fuzzy measure theory. In fact, non-additive measure has been used with various names in various fields: J. von Neumann and O. Morgenstern [52] called cooperative games in economics, which is without integral. Auman and Shapley [1] organized the theory of non atomic games systematically. Choquet[6] studied a non additive set function

V. Torra, Y. Narukawa, and M. Sugeno (eds.), *Non-Additive Measures*,
Studies in Fuzziness and Soft Computing 310,
DOI: 10.1007/978-3-319-03155-2_2,

called capacity on the class of compact sets in potential theory. He also considered a functional extending capacity, which was called Choquet integral later. Sugeno[65] called a non additive set function a fuzzy measure and proposed the integral with respect to a fuzzy measure in systems science. He named his integral the fuzzy integral, which is known as Sugeno integral today. Schmeidler [61] used a non additive set function, named a non additive subjective probability in Economics and he considered the Choquet integral with respect to a non additive subjective probability. The researchers of Discrete convex analysis study a non additive set function called matroid or submodular function [40,41,14]. Their Loväsz extension is the same as the Choquet integral. Denneberg [10] 's monograph titled non additive measure and integral show a lot of mathematical results.

In this paper, we present several basic definitions and results of non additive measures and integral. The paper is organized as follows: In section 2, we show some basic definition of a non additive measure. Comonotonicity is the important concept, considering the integral. Continuity of a non additive measure is useful for some convergence theorems of functions or integrals. We show some basic definitions of integral with respect to a non additive measure in Section 3. The most fundamental integrals are Sugeno integral and Choquet integral. We show their definition and basic properties. Some results about the extension of Choquet integral are presented in Section 3.

The generalization of both Sugeno and Choquet integral is called a generalized fuzzy integral in Section 4. We present a multidimensional integral which correspond to Fubini's theorem in classical theory.

In section 5, we study Möbius transform of a non additive measure and a representation of integral using the Möbius transform. We show that the generalized representation theorems of Choquet integral and show that they coincide if the universal set is finite.

We conclude with Concluding remarks.

2 Non-Additive Meeasure or Fuzzy Measure

Let S be a universal set and $\mathcal{S}$ be a sigma algebra of S. We say that $(S, \mathcal{S})$ is a measurable space. In this section, we will present several properties of non-additive measures.

2.1 Basic Definitions

Definition 1. Let $(S, \mathcal{S})$ be a measurable space and R_+ be a set of non-negative real numbers.

We say that a set function $\mu : \mathcal{S} \longrightarrow R_+$ is a non-additive measure or fuzzy measure if μ satisfies the next conditions:

1. $\mu(\emptyset) = 0$,
2. $A \subset B,\ A, B \in \mathcal{S} \Rightarrow \mu(A) \leq \mu(B)$.

Remark. If a set function μ on $\mathcal{S}$ satisfies $\mu(\emptyset) = 0$, we say that μ is a non monotonic non-additive measure or a non monotonic fuzz measure. A non monotonic fuzzy measure is studied in [36,75]

Example 1. (Old rare books) *Suppose there are rare books consisting of two volumes: x_1, x_2. Let $X := \{x_1, x_2\}$. Suppose that there is a secondhand bookseller who buys them at the prices: $v(\{x_1\})$ dollars per first volume, $v(\{x_2\})$ dollars per second volume, $v(X)$ dollars per set of two volumes. If he sets a high value on a complete set, then*

$$v(X) > v(\{x_1\}) + v(\{x_2\}).$$

Example 2. (The workers in a workshop)

Let X be the set of the workers in a workshop, and suppose that they produce the same products. For each $A \in 2^X$, the members of A work in the workshop. A group A may have various ways to work: Let $\mu(A)$ be the number of the products made by a group A in one hour. Then μ is a measure of the productivity of a group: the attribute of 2^X in question is the productivity. By the definition, the following statements are natural:

- *A and B work separately, then $\mu(A \cup B) = \mu(A) + \mu(B)$.*
- *A and B work with effective cooperation, $\mu(A \cup B) > \mu(A) + \mu(B)$.*
- *A and B work with incompatibility between A's operations and B's, $\mu(A \cup B) < \mu(A) + \mu(B)$*

As shown in the previous examples, a non additive measure represents a fact that can not be represent with an additive set function. The biggest problem is how to define each $\mu(A)$ for $A \in 2^X$. To avoid combinatorial explosion, a lot of special non-additive measures are proposed as follows.

Definition 2. *Let μ be a non-additive measure on $(S, \mathcal{S})$.*

1. *Sugeno λ measure: [65] for some fixed $\lambda > -1$ it holds*

$$\mu(A \cup B) = \mu(A) + \mu(B) + \lambda\mu(A)\mu(B)$$

   for all $A \cap B = \emptyset$
2. *Possibility measure:Zadeh [77] in the context of fuzzy sets.*

$$Pos(A \cup B) = \max(Pos(A), Pos(B))$$

3. *Necessity measure:[77]*

$$Nec(A \cap B) = \min(Nec(A), Nec(B))$$

4. *the 0-1 possibility measure Pos_A focused on A:*

$$\mathrm{Pos}_A(B) = 1 \text{ if } A \cap B \neq \emptyset, = 0 \text{ if } A \cap B = \emptyset$$

5. *the 0-1 necessity measure* Nec_A *focused on* A *(Unaminity game):*

$$\mathrm{Nec}_A(B) = 1 \text{ if } A \subseteq B, = 0 \text{ if } A \not\subset B$$

6. *[73,53]* μ *is called null-additive if* $\mu(A \cup B) = \mu(A)$ *whenever* $A, B \in \mathcal{S}$, $A \cap B = \emptyset$ *and* $\mu(B) = 0$.
7. *[74]* μ *is called weakly null-additive if* $\mu(A \cup B) = \mu(A)$ *whenever* $A, B \in \mathcal{S}$, $A \cap B = \emptyset$, $\mu(A) = 0$ *and* $\mu(B) = 0$.
8. *Decomposable measure: [76,34]*
if there exists a t-conorm $\perp$ *such that for all* $A, B \subseteq X$ *with* $A \cap B = \emptyset$ *it holds:*

$$\mu(A \cup B) = \mu(A) \perp \mu(B),$$

where $\perp$ *is t-conorm [18,30]* $\perp : [0,1] \times [0,1] \to [0,1]$ *:*
(i) $\perp(x, y) = \perp(y, x)$ *(commutativity)*
(ii) $\perp(\perp(x, y), z) = \perp(x, \perp(y, z))$ *(associativity)*
(iii) $\perp(x, y) \leq \perp(x', y')$ *if* $x \leq x'$ *and* $y \leq y'$ *(monotonicity)*
(iv) $\perp(x, 0) = x$ *for all* x *(neutral element 0)*
9. *Distorted probability: [54,4]*
$\mu(A) := f(P(A))$, *where* P *is probability and* f *is monotone function with* $f(0) = 0, f(1) = 1$.
10. *Belief function:[9,63]*

$$Bel(A_1 \cup ... \cup A_n) \geq \sum_j Bel(A_j) - \sum_{j<k} Bel(A_j \cap A_k) + ... \\ +(-1)^{n+1} Bel(A_1 \cap ... \cap A_n)$$

11. *Plausibility function: [9,63]*

$$Pl(A_1 \cap ... \cap A_n) \leq \sum_j Pl(A_j) - \sum_{j<k} Pl(A_j \cup A_k) + ... \\ +(-1)^{n+1} Pl(A_1 \cup ... \cup A_n)$$

12. *k order monotonicity: [6,5]*
(a) k-order monotone (or k-monotone) for $k \geq 2$, *if for all family of* k *subsets* $A_1, \ldots A_k$ *in* X,

$$\mu(\bigcup_{i=1}^{k} A_i) \geq \sum_{\emptyset \neq I \subset 1,\ldots k} (-1)^{|I|+1} \mu(\bigcap_{i \in A_i})$$

1-monotonicity is defined as monotonicity.
(b) totally monotone if it is k-monotone for any $k \geq 1$.
(c) k-order alternative (or k-alternative) for $k \geq 2$, *if k-order monotone (or k-monotone) if for all family of* k *subsets* $A_1, \ldots A_k$ *in* X,

$$\mu(\bigcap_{i=1}^{k} A_i) \leq \sum_{\emptyset \neq I \subset 1,\ldots k} (-1)^{|I|+1} \mu(\bigcup_{i \in A_i})$$

2-monotonicity is sometimes known by super modularity or convexity, 2-alternating fuzzy measures are sometimes called submodular measure.

13. *Inter-additive fuzzy measure: [35]*
 Let $\mathcal{P} = \{X_1, \dots, X_s\}$ be a partition of X then $\mathcal{P}$ is a μ-inter-additive partition of X if

$$\mu(A) = \sum_{X_i \in \mathcal{P}} \mu(A \cap X_i)$$

 for every $A \in 2^X$.

2.2 Measurable Function and Comonotonicity

Let $(S, \mathcal{S})$ be a measurable space. We say that a function $f : S \to R$ is measurable, if $\{x|f(x) \geq a\} \in \mathcal{S}$ for all real number a. $\mathcal{F}(S)$ denotes the class of measurable functions on S. and $\mathcal{F}^+(S)$ denotes the class of non negative measurable functions on S.

The concepts related to convergence of measurable functions are the same as the classical ones.

Definition 3. *Let $f, f_n \in \mathcal{F}(S)$ $(n = 1, 2, \dots)$.*

1. *We say that $\{f\}$ converges almost everywhere to f on S, and denote it by $f_n \to (\mu - a.e.)f$, if there exists a subset $E \subset S$ such that $\mu(E) = 0$ and $f_n \to f$ on $S \setminus E$.*
2. *We say that $\{f\}$ converges almost uniformly to f on S, and denote it by $f_n \to (\mu - a.u.)f$, if for any $\epsilon > 0$ there exists a subset $E_\epsilon \in \mathcal{S}$ such that $\mu(S \setminus E_\epsilon) < \epsilon$ and $f_n \to f$ uniformly on E_ϵ.*

The comonotonicity of measurable functions are one of the most important properties when we consider the additivity of integral.

Definition 4. $f, g \in \mathcal{F}(S)$ f and g are comonotonic, if $x, x' \in S$

$$f(x) < f(x') \Rightarrow g(x) \leq g(x')$$

The next proposition gives a necessary and sufficient condition for comonotonicity.

Proposition 5. [10] *Let $f, g \in \mathcal{F}(S)$. The following conditions are equivalent.*

1. *f and g are comonotonic.*
2. *$(f(x_1) - f(x_2))(g(x_1) - g(x_2)) \leq 0$ for all $x_1, x_2 \in S$.*
3. *$\{x|f(x) \geq a\} \subset \{x|g(x) \geq b\}$ or $\{x|f(x) \geq a\} \supset \{x|g(x) \geq b\}$ for all $a, b \in R$.*
4. *There exists a function $F : S \to R$ and increasing functions u, v on R such that $f = u(F), g = v(F)$.*
5. *There exist continuous increasing functions u, v on R such that $u(z) + v(z) = z, z \in R$ and $f = u(f + g), g = v(f + g)$.*

2.3 Continuity

The continuity is a basic assumption for the convergence theorem of the sequence of measurable functions. Sugeno's original definition of fuzzy measure [65] assumes continuity as shown in Definition 6. However it is pointed out that Sugeno's continuity is too restrictive. That is, Possibility measure and Necessity measure are not included in Fuzzy measure [55]. Due to this problem, several alternative definitions of continuity have been proposed. Definition 7 reviews some of them.

Definition 6. *Let μ be a non additive measure on $(S, \mathcal{S})$ and $A, A_n \in \mathcal{S}$ $(n = 1, 2, \dots)$.*

1. *μ is said to be continuous from below if for every increasing sequence $\{A_n\}$ of measurable sets, it holds that*
$$\mu(\lim_{n\to\infty} A_n) = \lim_{n\to\infty} \mu(A_n),$$
that is, $A_n \uparrow A \Rightarrow \mu(A_n) \uparrow \mu(A)$.
2. *μ is said to be continuous from above if for every decreasing sequence $\{A_n\}$ of measurable sets, it holds that*
$$\mu(\lim_{n\to\infty} A_n) = \lim_{n\to\infty} \mu(A_n),$$
that is, $A_n \downarrow A \Rightarrow \mu(A_n) \downarrow \mu(A)$.
3. *We say that a non additive measure which is continuous from both above and below is continuous.*

Definition 7. *Let μ be a non additive measure on $(S, \mathcal{S})$.*

1. *[73] μ is said to be autocontinuous from above if for $A \in \mathcal{B}$ $\lim_{n\to\infty} \mu(A \cup B_n) = \mu(A)$ whenever $\lim_{n\to\infty} \mu(B_n) = 0$.*
2. *[22] μ is said to be strongly order continuous if $N_n \downarrow N$ and $\mu(N) = 0$ imply $\mu(Nn) \downarrow 0$.*
3. *[39] μ is said to be strongly order totally continuous if, for every decreasing net $\mathcal{B}$ of measurable sets such that $\cap\mathcal{B}$ is measurable and $\mu(\cap\mathcal{B}) = 0$, $\inf_{B\in\mathcal{B}} \mu(B) = 0$.*
4. *[71] μ is said to be null-continuous if $N_n \uparrow N$ and $\mu(Nn) = 0$ for every $n = 1, 2, \dots$ imply $\mu(N) = 0$.*
5. *[67] μ said to have property (S) if $\mu(N_n) \to 0$ implies that there exists a subsequence $\{N_{n_i}\}$ of $\{N_n\}$ such that $\mu(\cap_{k=1}^{\infty} \cup_{i=k}^{\infty} N_{n_i}) = 0$.*
6. *[39] μ is said to satisfy the Egoroff condition (for short Ec) if, for every doubly-indexed sequence $N_{m,n}$ such that $N_{m,n} \supset N_{m',n'}$ for $m \geq m'$ and $n \leq n'$ and $\mu(\cup_{m=1}^{\infty} \cap_{n=1}^{\infty} N_{m,n}) = 0$, and for every $\epsilon > 0$, there exists a sequence $\{n_m\}$ such that $\mu(\cup_{m=1}^{\infty} N_{m,n_m}) < \epsilon$.*
7. *[12] We say that μ has a pseudo metric generating property (for short "p.g.p."), if both $\lim_{n\to\infty} \mu(A_n) = 0$ and $\lim_{n\to\infty} \mu(B_n) = 0$ implies $\lim_{n\to\infty} \mu(A_n \cup B_n) = 0$ for $\{A_n\}$, $\{B_n\} \subset \mathcal{B}$.*

8. *[13] μ is said to be exhaustive if* $\lim_{n\to\infty}\mu(A_n) = 0$ *for any infinite disjoint sequence* $\{A_n\} \subset \mathcal{B}$.
9. *[23] μ is said to satisfy condition (E) if* $N_n^m \downarrow N^m$ *as* $n \to \infty$ *for every* m *and* $\mu(\cup_{m=1}^{\infty} N^m) = 0$ *imply that there exist strictly increasing sequences* $\{n_i\}$ *and* $\{m_i\}$ *such that* $\mu(\cup_{i=k}^{\infty} N_{n_i}^{m_i}) \to 0$ *as* $k \to \infty$.
10. *[68] μ is said to satisfy condition (M) if* $\mu(\cup_{n=1}^{\infty}\cap_{i=n}^{\infty} N_i) = 0$ *implies that for every* ϵ *there exists a strictly increasing sequence* $\{m_n\}$ *such that* $\mu(\cup_{n=1}^{\infty}\cap_{i=n}^{m_n} N_i) < \epsilon$.

Theorem 8. *(Egoroff's theorem) [23,39] Let μ be a monotone non-additive measure on $\mathcal{S}$. The next conditions are equivalent.*

1. *μ satisfies condition (Ec).*
2. *μ satisfies condition (E).*
3. $f_n \to^{(\mu-a.e.)} f$ *implies* $f_n \to^{(\mu-a.u.)} f$.

The interrelations among each definition are summarized as follows.

Theorem 9. *Let μ be a monotone non-additive measure on $\mathcal{S}$.*

1. *[24] Continuity implies condition (E).*
2. *[23,39] Strong order continuity and property (S) implies condition (E).*
3. *[39] Strong order total continuity implies condition (E).*
4. *[68] Continuity implies condition (M) and null-continuity.*
5. *[68] Strong order continuity and property (S) implies condition (M) and null-continuity.*
6. *[23,39] Condition (E) implies strong order continuity.*
7. *[68] Condition (M) and null-continuity implies condition (E).*
8. *[25] Strongly order continuity and property (S) implies property (E).*
9. *[25] If μ is order continuous and has p.g.p., then it is strongly order continuous and has property (S).*

3 Integral with Respect to a Non Additive Measure

In this section, we will present the integral with respect to a non-additive measure, that is, Sugeno integral and Choquet integral.

3.1 Sugeno Integral

Sugeno [65] proposed the integral with respect to a non additive measure. The integral was called a fuzzy integral. However it is called Sugeno integral now.

Definition 10. Let μ be a non-additive measure on $(S, \mathcal{S})$ with $\mu(S) = 1$

Let $f : S \to [0, 1]$ be a measurable. The Sugeno integral of f with respect to μ is defined as follows;

$$(S)\int f d\mu := \vee_{r\in[0,1]}(r \wedge \mu_f(r))$$

where $\mu_f(r) := \mu(\{x|f(x) > r\})$.

The Sugeno integral is shown as the coordinate of intersecting point of the graph below

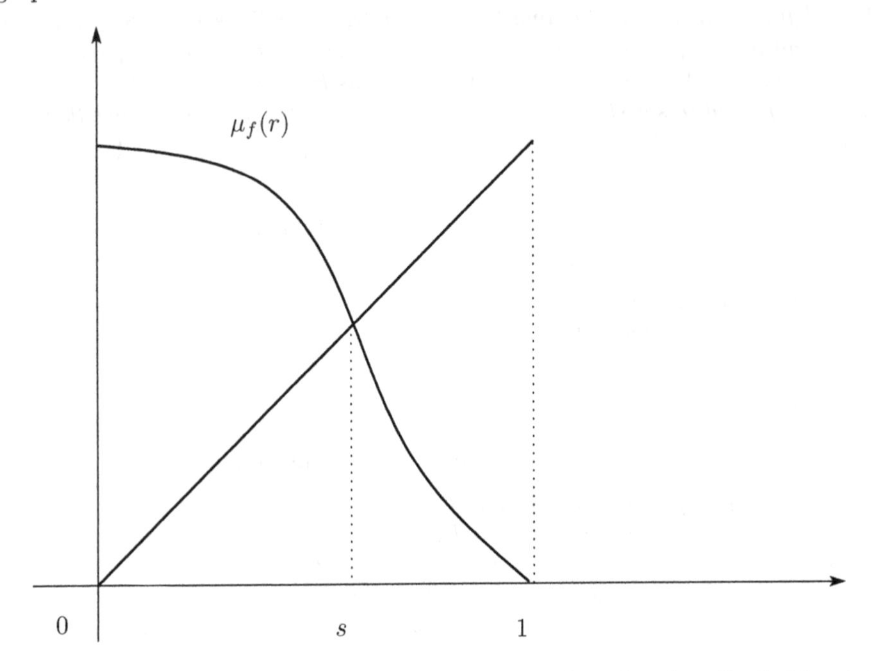

The integrand f for Sugeno integral is restricted to $f : S \to [0, 1]$. The definition of Sugeno integral can be extend so that its integrand is a member of $\mathcal{F}^+(S)$.

Definition 11. Let μ be a non-additive measure on $(S, \mathcal{S})$ with $\mu(S) = \infty$. Let $f \in \mathcal{F}^+(S)$. The Sugeno integral of f with respect to μ is defined as

$$(S)\int f d\mu := \vee_{r\in[0,\infty)}(r \wedge \mu_f(r))$$

where $\mu_f(r) := \mu(\{x|f(x) > r\})$.

The basic properties of Sugeno integral is following:

Theorem 12. *Let μ be a non-additive measure on $(S, \mathcal{S})$ with $\mu(S) = \infty(resp.1)$.*

1. $(S)\int a \wedge f d\mu = a \wedge (S)\int f d\mu$ *for* $f, g \in \mathcal{F}^+(S)(resp.with 0 \leq f, g \leq 1)$
2. $f \leq g$ *implies* $(S)\int f d\mu \leq (S)\int g d\mu$
3. $(S)\int f \vee g d\mu = (S)\int f \vee g d\mu$ *for comonotonic* $f, g \in \mathcal{F}^+(S)$.

Suppose that a non additive measure μ is continuous. Then the following monotone convergence theorem for Sugeno integral can be shown [65].

Theorem 13. *[65] Let μ be a continuous non-additive measure on $(S, \mathcal{S})$ with $\mu(S) = \infty$. Let $\{f_n\}$ be a monotone sequence of measurable functions.*

Then we have

$$\lim_{n\to\infty}(S)\int f_n d\mu = (S)\int \lim_{n\to\infty} f_n d\mu.$$

Sugeno integral uses only the order of each value of the function, and does not require addition or multiplication. Sugeno integral is useful for problem with an ordinal scale, but it is weak for transformation of scale. Even if μ is $\sigma-$additive, Sugeno integral des not coincide with Lebesgue integral. That is, Sugeno integral is not an extension of classical integral.

3.2 Choquet Integral

Choquet [6] proposed his functional in potential theory, which is an extension of capacity on a class of compact sets to a sets of functions. Later the functional was regarded as the integral with respect to a non additive measure [32,61]. Vitali [72] introduced a similar idea on 1925, although he did not consider a non additive measure. König [20] proposed horizontal integral, which is also similar to the Choquet idea.

Definition 14. Let μ be a non additive measure on $(S, \mathcal{S})$ and $f : S \to [0, \infty)$. The Choquet integral of f with respect to μ is defined by

$$(C)\int f d\mu := \int_0^\infty \mu_f(r)dr,$$

where $\mu_f(r) := \mu(\{x | f(x) > r\})$.

Example 3. *(Old rare books) Suppose that there is a secondhand bookseller who buys them at the prices: $v(\{x_1\})$ dollars per first volume, $v(\{x_2\})$ dollars per second volume, $v(X)$ dollars per set of two volumes.*

Since he sets a high value on a complete set,

$$v(X) > v(\{x_1\}) + v(\{x_2\}).$$

A certain person sells $f(x_1)$ first volumes and $f(x_2)$ second volumes to the secondhand bookseller. We may assume that $f(x_1) > f(x_2)$.

He gets

$$f(x_2)\mu(X) + (f(x_1) - f(x_2))\mu(x_1).$$

This is a Choquet integral.

Example 4. *(The workers in a workshop)*

Let X be the set of the workers in a workshop, and suppose that they produce the same products. For each $A \in 2^X$, the members of A work in the workshop. Let $\mu(A)$ be the number of the products made by A in one hour. Suppose a person x_i works $f(x_i)$ hour and $f(x_1) \leq f(x_2) \leq \cdots \leq f(x_n)$.

The group (all member) X works $f(x_1)$ hours. the group (member excludes x_1) $X \setminus \{x_1\}$ works $f(x_2) - f(x_1)$ hours. the groups $X \setminus \{x_1, x_2\}$ works $f(x_3) - f(x_2)$ hours. and so on.

Last x_n works $f(x_n) - f(x_{n-1})$ hours.

Then the total numbers of the products is

$$f(x_1)\mu(X) + (f(x_2) - f(x_1))\mu(X \setminus \{x_1\}) + (f(x_3) - f(x_2))\mu(X \setminus \{x_1, x_2\}) \\ + \cdots + (f(x_n) - f(x_{n-1})\mu(\{x_n\})$$

This is the Choquet integral of f with respect to μ.

The basic properties of Choquet integral are presented as follows.

Theorem 15. *Let μ be a non-additive measure on $(S, \mathcal{S})$.*

1. $(C)\int 1_A d\mu = a\mu(A)$ *where 1_A is a characteristic of $A \in \mathcal{S}$.*
2. $(C)\int af d\mu = a(C)\int f d\mu$ *for $f \in \mathcal{F}^+(S)$.*
3. $f \le g$, $f, g \in \mathcal{F}^+(S)$. *implies* $(C)\int f d\mu \le (C)\int g d\mu$.

Suppose that a non additive measure μ is continuous. Then we have the monotone convergence theorem for the Choquet integral. The theorem follows and its proof is in [10].

Theorem 16. *Let μ be a non-additive measure on $(S, \mathcal{S})$, which is continuous from below. Let $\{f_n\}$ be a monotone increasing sequence of measurable functions.*

Then we have

$$\lim_{n\to\infty} (C)\int f_n d\mu = (C)\int \lim_{n\to\infty} f_n d\mu.$$

Generally the Choquet integral is not additive. We can the consider the condition under which the Choquet integral is additive. Dellacherie [8] shows that the Choquet integral is additive if f and g are comonotonic.

Theorem 17. *[8,10] If $f, g \in \mathcal{F}^+(S)$ are comonotonic, then*

$$(C)\int f + g d\mu = (C)\int f d\mu + (C)\int g d\mu.$$

Taking a notice of a measure, we have the next theorem from the definition.

Theorem 18. *Let $(S, \mathcal{S})$ be a measurable space, f be a bounded nonnegative measurable function on S and $A \in \mathcal{S}$.*

1. $(C)\int f d\mathrm{Pos}_A = \max_{x \in A} f(x)$ *where Pos_A is the 0-1 possibility measure focused on A.*

2. $(C)\int f d\mathrm{Nec}_A = \min_{x\in A} f(x)$ *where* Nec_A *is the 0-1 necessity measure focused on* A.

If we consider the Sugeno integral, we have a similar max-min theorem as the one for the Choquet integral.

Theorem 19. *Let* $(S,\mathcal{S})$ *be a measurable space,* f *be a bounded nonnegative measurable function on* S *with* $0 \le f \le 1$ *and* $A \in \mathcal{S}$.

1. $(S)\int f d\mathrm{Pos}_A = \max_{x\in A} f(x)$ *where* Pos_A *is the 0-1 possibility measure focused on* A.
2. $(S)\int f d\mathrm{Nec}_A = \min_{x\in A} f(x)$ *where* Nec_A *is the 0-1 necessity measure focused on* A.

3.3 Extension of Choquet Integral

We have defined a Choquet integral of a non-negative function with respect to a non additive measure. In this subsection, we will show some extensions of the Choquet integral for a real valued function f that may take a negative value.

If $\mu(S) < \infty$, we can define the Choquet integral of f.

Definition 20. Suppose that $(S,\mathcal{S})$ be a measurable space and μ be a non additive measure on $\mathcal{S}$ with $\mu(S) < \infty$.

Let $f \in \mathcal{F}(S)$.

$$(C)\int f\mu := (C)\int (f \vee 0)\mu - (C)\int ((-f)\vee 0) d\bar{\mu}$$

where $\bar{\mu}(A) := \mu(S) - \mu(A)$ for $A \in \mathcal{S}$.

We say that $\bar{\mu}$ is a conjugate of μ. Let $f \in \mathcal{F}(S)$, $C_\mu(f)$ denotes a Choquet integral of f with respect to μ, that is,

$$C_\mu(f) = (C)\int f d\mu.$$

This integral is known as the asymmetric integral.

The extended Choquet integral has both monotonicity and comonotonic additivity. Conversely, a functional on $\mathcal{F}(S)$ with both monotonicity and comonotonic additivity can be represented as a Choquet integral.

Theorem 21. *[10,61]*

1. $f, g \in \mathcal{F}(S)$ *are comonotonic,*

$$C_\mu(f+g) = C_\mu(f) + C_\mu(g).$$

2. *Let* $I : \mathcal{F}(S) \to R$ *with* $I(f) \leq I(g)$ *for* $f \leq g$ $f, g \in \mathcal{F}(S)$ *and* $I(f+g) = I(f) + I(g)$ *for comonotonic* $f, g \in \mathcal{F}(S)$. *Then there exists a non-additive measure such that*
$$I(f) = C_\mu(f)$$
for $f \in \mathcal{F}(S)$.

Another extension is known as Šipoš integral.

Definition 22. [64]. Suppose that $(S, \mathcal{S})$ be a measurable space and μ be a non additive measure on $\mathcal{S}$.

A Šipoš integral or symmetric integral is defined by

$$C_\mu(f) := C_\mu(f \vee 0) - C_\mu((-f) \vee 0).$$

for $f \in \mathcal{F}(S)$.

The Cumulative Prospect Theory (CPT) by Tversky and Kahnemann [70] is a generalization of both the asymmetric integral and the Šipoš integral. We review CPT below.

Definition 23. Suppose that $(S, \mathcal{S})$ be a measurable space and μ^+ and μ^- be a non additive measure on $\mathcal{S}$. A CPT functional $CPT : \mathcal{F}(S) \to R$ is defined by

$$CPT_{\mu^+,\mu^-}(f) := C_{\mu^+}(f \vee 0) - C_{\mu^-}((-f) \vee 0).$$

The next theorem presents a necessary and sufficient condition for a CPT functional on some topological setting.

Theorem 24. [43] *Let* S *be a Locally compact Hausdorff space and* $C(S)$ *be a class of continuous functions with compact support and* $I : C(S) \to R$ *a comonotonically additive and monotone functional.*

Then there exist non additive measures μ^+, μ^- *on* $(S, \mathcal{S})$ *such that*

$$I(f) = CPT_{\mu^+,\mu^-}(f)$$

for $f \in C(S)$.

If S *is compact, we have* $\mu^- = \overline{\mu^+}$.

Example 5. *Let* S *be a countable set, that is,* $S := \{a_1, a_2, \ldots\}$. *Then* S *is a locally compact Hausdorff space with the discrete topology and* $\mathcal{S} = 2^S$. *Let* $C(S)$ *be the class of continuous functions with compact support and* $f \in C(S)$. *Then* $|f(S)| < \infty$, *that is, the support of* f *is a finite set. Let* $I : C(S) \to R$ *a comonotonically additive and monotone functional.*

Then there exist non additive measures μ^+, μ^- *on* $(S, \mathcal{S})$ *such that*

$$I(f) = CPT_{\mu^+,\mu^-}(f)$$

for $f \in C(S)$.

If S is a finite set, that is, $S := \{a_1, a_2, \ldots, a_n\}$. $C(S)$ is a class of all real valued function on S. Since $\mu^- = \overline{\mu^+}$, we have

$$CPT_{\mu^+,\mu^-}(f) = (C)\int f d\mu^+$$

for $f \in C(S)$.

Remark. Let L be a lattice. A lattice valued Sugeno integral is proposed and studied in [16], where a CPT type extension of Sugeno integral is also proposed and studied. The representation of L valued functional as Sugeno integral is also studied in [58,7,29].

Next we will show two other extensions of Choquet integral, that is, the Choquet integral with constant and the multistep Choquet integral.

Definition 25. *Let $(S, \mathcal{S})$ be a measurable space, μ a non additive measure on $\mathcal{S}$ and b is a real number. The Choquet integral with constant b of $f : S \to R^+$ with respect to μ is defined by*
$C_{\mu,b}(f) = C_\mu(f) + b.$

The multistep Choquet integral is proposed and studied in [31,37,38,44]

Definition 26. *Let S be a finite set and $\mathcal{S} = 2^S$.*

Let $\mu_1, \mu_2, \ldots, \mu_n$ be non additive measures on $\mathcal{S}$ and $b_1, b_2, \ldots, b_n$ be real numbers. Then we define the 1st step Choquet integrals as a set of Choquet integrals with constant: $\{C_{\mu_1,b_1}(f), \ldots, C_{\mu_n,b_n}(f)\}$ for $f : S \to R^+$.

Let μ be a non additive measure on $2^{1,\ldots,n}$ and b a real number. Two step Choquet integral I of f with constant is defined by $I(f) := C_{\mu,b}(C_{\mu_i,b_i})$.

We can define the $n-$step integral with constant inductively.

Theorem 27. *[44]*

Sugeno integral can be represented as a two-step Choquet integral with constant.

4 Generalized Fuzzy Integral

We define a generalized non additive integral in terms of a pseudo-addition $\oplus$ and a pseudo-multiplication $\boxdot$. The generalized non additive integral is one of the generalizations of both Choquet integral and Sugeno integral. Formally, $\oplus$ and $\boxdot$ are binary operators that generalize addition and multiplication, and also max and min. We want to recall that generalized fuzzy integrals have been investigated by Benvenuti et al. in [2].

4.1 Generalized Fuzzy Integral

Note that we will use $k \in (0, \infty)$ in the rest of this paper.

Definition 28. *A pseudo-addition $\oplus$ is a binary operation on $[0,k]$ or $[0,\infty)$ fulfilling the following conditions:*

(A1) $x \oplus 0 = 0 \oplus x = x$.
(A2) $x \oplus y \leq u \oplus v$ *whenever* $x \leq u$ *and* $y \leq v$.
(A3) $x \oplus y = y \oplus x$.
(A4) $(x \oplus y) \oplus z = x \oplus (y \oplus z)$.
(A5) $x_n \to x, y_n \to y$ *implies* $x_n \oplus y_n \to x \oplus y$.

A pseudo-addition $\oplus$ is said to be strict if and only if $x \oplus y < x \oplus z$ whenever $x > 0$ and $y < z$, for $x, y, z \in (0,k)$; and it is said to be Archimedean if and only if $x \oplus x > x$ for all $x \in (0,k)$.

Definition 29. *A pseudo-multiplication $\boxdot$ is a binary operation on $[0,k]$ or $[0,\infty)$ fulfilling the conditions:*

(M1) *There exists a unit element $e \in (0,k]$ such that $x \boxdot e = e \boxdot x = x$.*
(M2) $x \boxdot y \leq u \boxdot v$ *whenever* $x \leq u$ *and* $y \leq v$.
(M3) $x \boxdot y = y \boxdot x$.
(M4) $(x \boxdot y) \boxdot z = x \boxdot (y \boxdot z)$.
(M5) $x_n \to x, y_n \to y$ *implies* $x_n \boxdot y_n \to x \boxdot y$.

Example 6.

1. The maximum operator $x \vee y$ is a non Archimedean pseudo-addition on $[0,k]$.
2. The sum $x + y$ is an Archimedean pseudo-addition on $[0,\infty)$.
3. The Sugeno operator $x +_\lambda y := 1 \wedge (x + y + \lambda xy)$ $(-1 < \lambda < \infty)$ is an Archimedean pseudo-addition on $[0,1]$.

Proposition 30. *(Ling, 1965, [27])*
If a pseudo-addition $\oplus$ is Archimedean, then there exists a continuous and strictly increasing function $g : [0,k] \to [0,\infty]$ such that $x \oplus y = g^{(-1)}(g(x) + g(y))$, where $g^{(-1)}$ is the pseudo-inverse of g defined by

$$g^{(-1)}(u) := \begin{cases} g^{(-1)}(u) & \text{if } u \leq g(k) \\ k & \text{if } u > g(k). \end{cases}$$

The function g is called an additive generator of $\oplus$.

Definition 31. *Let μ be a non additive measure on a fuzzy measurable space $(S, \mathcal{S})$; then, we say that μ is a $\oplus$-measure or a $\oplus$-decomposable non additive measure if $\mu(A \cup B) = \mu(A) \oplus \mu(B)$ whenever $A \cap B = \emptyset$ for $A, B \in \mathcal{S}$.*

A $\oplus$-measure μ is called normal when either $\oplus = \vee$, or $\oplus$ is Archimedean and $g \circ \mu$ is an additive measure. Here, g corresponds to an additive generator of $\oplus$.

When $\oplus$ is a t-conorm $\perp$, the $\oplus$-measure μ is the decomposable measure in Definition 2 (8).

Definition 32. *Let $k \in (0, \infty)$, let $\oplus$ be a pseudo-addition on $[0, k]$ or $[0, \infty)$ and let $\boxdot$ be a pseudo-multiplication on $[0, k]$ or $[0, \infty)$; then, we say that $\boxdot$ is $\oplus$-fitting if*

(F1) $a \boxdot x = 0$ *implies* $a = 0$ *or* $x = 0$, *and* $a \downarrow 0$ *implies* $a \boxdot x \downarrow 0$.
(F2) $a \boxdot (x \oplus y) = (a \boxdot x) \oplus (a \boxdot y)$.
(F3) $(a \oplus b) \boxdot x = (a \boxdot x) \oplus (b \boxdot x)$.

Under these conditions, we say that $(\oplus, \boxdot)$ is a pseudo-fitting system.

Let $\oplus$ be a pseudo-addition; then, we define its pseudo-inverse $-_{\oplus}$ as

$$a -_{\oplus} b := \inf\{c | b \oplus c \geq a\}$$

for all $(a, b) \in [0, k]^2$.

Definition 33. *[2] For any $r > 0$ and $A \in \mathcal{S}$, the basic simple function $b(r, A)$ is defined by $b(r, A)(x) = r$ if $x \in A$ and $b(r, A)(x) = 0$ if $x \notin A$.*

Then, we say that a function f is a simple function if it can be expressed as

$$f := \sum_{i=1}^{n} b(a_i, A_i) \text{ for } a_i > 0 \tag{1}$$

where $A_1 \supsetneq A_2 \supsetneq \cdots \supsetneq A_n$, $A_i \in \mathcal{S}$.

Expression 1 is called a comonotonic additive representation of f. f can also be expressed as $f := \vee_{i=1}^{n} b(a'_i, A_i)$ for $a'_1 > \cdots > a'_n > 0$, where $A_1 \supsetneq A_2 \supsetneq \cdots \supsetneq A_n$, $A_i \in \mathcal{S}$. This expression is called a comonotonic maxitive representation of f.

Definition 34. [66] *Let μ be a non additive measure on a measurable space $(X, \mathcal{X})$, and let $(\oplus, \boxdot)$ be a pseudo-fitting system. Then, when μ is a normal $\oplus$-measure, we define the pseudo-decomposable integral of a measurable simple function f on X such that $f = \oplus 1_{i=1}^{n} b(r_i, D_i)$ where $D_i \cap D_j \neq \emptyset$ for $i \neq j$, as follows:*

$$(D) \int f d\mu := \oplus_{i=1}^{n} r_i \boxdot \mu(D_i).$$

Since μ is an $\oplus$-measure, it is obvious that the integral is well defined.

Definition 35. *Let μ be a non additive measure on a measurable space $(S, \mathcal{S})$, and let $(\oplus, \boxdot)$ be a pseudo-fitting system. Then, the generalized fuzzy integral (GF-integral) of a measurable simple function $f := \oplus_{i=1}^{n} b(a_i, A_i)$, with $a_i > 0$ and $A_1 \supsetneq A_2 \supsetneq \ldots A_n$, $A_i \in \mathcal{X}$, is defined as follows:*

$$(GF) \int f d\mu := \oplus_{i=1}^{n} a_i \boxdot \mu(A_i).$$

The GF-integral of a simple function is well defined [2].

The next proposition follows from the definition of the pseudo-inverse $-_{\oplus}$, the generalized t-conorm integral (Definition 35), and the t-conorm integral (Definition 34).

Proposition 36. *Let μ be a non additive measure on a measurable space $(S, \mathcal{S})$, and let $(\oplus, \boxdot)$ be a pseudo-fitting system. Then, if μ is a normal $\oplus$-measure, the generalized fuzzy integral coincides with the pseudo-decomposable integral.*

Example 7

1. When $\oplus = +$ and $\boxdot = \cdot$, the generalized fuzzy integral is a Choquet integral.
2. When $\oplus = \vee$ and $\boxdot = \wedge$, the generalized fuzzy integral is a Sugeno integral.

Theorem 37. *[49] Let $(S, \mathcal{S}, \mu)$ be a fuzzy measure space and let $(\oplus, \boxdot)$ be a pseudo-fitting system. Then, for comonotonic measurable functions f, and g, we have*

$$(GF)\int(f \oplus g)d\mu = (GF)\int f d\mu \oplus (GF)\int g d\mu.$$

We call this property the comonotonic $\oplus$-additivity of a generalized fuzzy integral.

4.2 Multidimensional Integral

We consider first the case of the product of two fuzzy measurable spaces. Let X and Y be two universal sets and $X \times Y$ be the direct product of X and Y, let $(X, \mathcal{X})$ and $(Y, \mathcal{Y})$ be two fuzzy measurable spaces; then, we define the following class of sets:

$$\mathcal{X} \times \mathcal{Y} := \{A \times B | A \in \mathcal{X}, B \in \mathcal{Y}\}$$

Now, let us consider the measurable space $(X \times Y, \mathcal{X} \times \mathcal{Y})$. Suppose that $\mathcal{X} := 2^X$ and $\mathcal{Y} := 2^Y$. Note that $\mathcal{X} \times \mathcal{Y} \neq 2^{X \times Y}$ if $|X| > 1$ and $|Y| > 1$.

Therefore, the class of $\mathcal{X} \times \mathcal{Y}$-measurable functions is smaller than the class of $2^{X \times Y}$-measurable functions.

Example 8. Let $X := \{x_1, x_2\}$ and $Y := \{y_1, y_2\}$; then, we have

$$\begin{aligned} 2^X \times 2^Y := \{ & \emptyset, \{(x_1, y_1)\}, \{(x_1, y_2)\}, \{(x_2, y_1)\}, \\ & \{(x_2, y_2)\}, \{(x_1, y_1), (x_2, y_1)\}, \{(x_1, y_2), (x_2, y_2)\}, \\ & \{(x_1, y_1), (x_1, y_2)\}, \{(x_2, y_1), (x_2, y_2)\}, \\ & \{(x_1, y_1), (x_1, y_2), (x_2, y_2), (x_2, y_2)\} \} \end{aligned}$$

Hence, $\{(x_1, y_1), (x_2, y_2)\} \notin 2^X \times 2^Y$.

The next proposition follows from the definition of a $\mathcal{X} \times \mathcal{Y}$-measurable function.

Proposition 38. *Let $f : X \times Y \to [0, k]$ be a $\mathcal{X} \times \mathcal{Y}$-measurable function; then,*

1. *for fixed $y \in Y$, $f(\cdot, y)$ is $\mathcal{X}$-measurable, and*
2. *for fixed $x \in X$, $f(x, \cdot)$ is $\mathcal{Y}$-measurable.*

Example 9. *Let $X := \{x_1, x_2\}$ and $Y := \{y_1, y_2\}$. Let $(X, \mathcal{X})$ and $(Y, \mathcal{Y})$ be the two measurable spaces defined with $\mathcal{X} = 2^X$ and $\mathcal{Y} = 2^Y$. Under these conditions, we consider the two functions below and study whether they are $\mathcal{X} \times \mathcal{Y}$-measurable functions.*

1. *Let us define $f : X \times Y \to [0, 1]$ by*
   $f(x_1, y_1) = f(x_1, y_2) = 0.2,$
   $f(x_2, y_1) = 0.6,$
   $f(x_2, y_2) = 1.$
   Then, we have,
   $$\begin{aligned} \{(x, y) | f(x, y) \geq 1\} &= \{(x_2, y_2)\} \\ &= \{x_2\} \times \{y_2\}, \\ \{(x, y) | f(x, y) \geq 0.6\} &= \{(x_2, y_1), (x_2, y_2)\} \\ &= \{x_2\} \times \{x_1, y_2\}, \\ \{(x, y) | f(x, y) \geq 0.2\} &= \{(x_1, y_1), (x_1, y_2), (x_2, y_1), (x_2, y_2)\} \\ &= \{x_1, x_2\} \times \{y_1, y_2\}. \end{aligned}$$
   Therefore, f is $\mathcal{X} \times \mathcal{Y}$-measurable.
2. *Let us define $g : X \times Y \to [0, 1]$ by*
   $g(x_1, y_1) = 0.2,$
   $g(x_1, y_2) = 0.4,$
   $g(x_2, y_1) = 0.6,$
   $g(x_2, y_2) = 1.$
   Then, we have
   $\{(x, y) | g(x, y) \geq 0.4\} = \{(x_1, x_2), (x_2, y_1), (x_2, y_2)\} \notin \mathcal{X} \times \mathcal{Y}.$
   Therefore, g is not a $\mathcal{X} \times \mathcal{Y}$-measurable function.

In fact, if $A \in \mathcal{X} \times \mathcal{Y}$, we have $|A| = 0, 1, 2,$ or 4.

Theorem 39. (Fubini-like theorem) [49] *Let $(X, \mathcal{X}, \mu)$ and $(Y, \mathcal{Y}, \nu)$ be two fuzzy measure spaces, $(\oplus, \boxdot)$ be a pseudo-fitting system and let $f : X \times Y \to [0, k]$ be a $\mathcal{X} \times \mathcal{Y}$-measurable function. Then, there exists a fuzzy measure m on $\mathcal{X} \times \mathcal{Y}$ such that*

$$\begin{aligned} (GF) \int ((GF) \int f d\mu) d\nu &= (GF) \int f dm \\ &= (GF) \int ((GF) \int f d\nu) d\mu. \end{aligned}$$

4.3 Extension of the Domain

We assume that $\mathcal{X}$ and $\mathcal{Y}$ are algebras. Even if $|X|$, $|Y|$ are finite and such that $\mathcal{X} := 2^X$ and $\mathcal{Y} := 2^Y$, the class of $\mathcal{X} \times \mathcal{Y}$ is smaller than the class of $2^{X \times Y}$, as we have shown in Example 21. In general, the class of $\mathcal{X} \times \mathcal{Y}$-measurable functions is too small as shown in Example 23.

We consider an extension of the domain of the measure. In general, unless there are additional constraints or conditions on the fuzzy measures, it is impossible to extend the domain. However, in our case, we assume that μ on $(X, \mathcal{X})$ and ν on $(Y, \mathcal{Y})$ are normal $\oplus$-measures. In this case, an extension is possible. We define the extension below.

Definition 40. *Let us define the class $\overline{\mathcal{X} \times \mathcal{Y}}$ of sets $A \in 2^{X \times Y}$ by*

$$\overline{\mathcal{X} \times \mathcal{Y}} := \{A \in 2^{X \times Y} | A = \cup_{i \in I} A_i, A_i \in \mathcal{X} \times \mathcal{Y}, I : \textit{finite}\}.$$

We say that $(X \times Y, \overline{\mathcal{X} \times \mathcal{Y}})$ is an extended fuzzy measurable space.

Proposition 41. *Let $(X, \mathcal{X})$ and $(Y, \mathcal{Y})$ be two fuzzy measurable spaces. Let $(X \times Y, \overline{\mathcal{X} \times \mathcal{Y}})$ be an extended fuzzy measurable space.*

1. *$\overline{\mathcal{X} \times \mathcal{Y}}$ is an algebra.*
2. *Let f on $X \times Y$ be $\overline{\mathcal{X} \times \mathcal{Y}}$-measurable*
   (a) *$f(x, \cdot)$ is $\mathcal{X}$-measurable.*
   (b) *$f(\cdot, y)$ is $\mathcal{Y}$-measurable.*

Corollary 42. *Let $(X \times Y, \overline{\mathcal{X} \times \mathcal{Y}})$ be an extended fuzzy measurable space. Let us suppose that $|X|$, $|Y|$ are finite. If $\mathcal{X} = 2^X$ and $\mathcal{Y} = 2^Y$, then $\overline{\mathcal{X} \times \mathcal{Y}} = 2^{X \times Y}$.*

It follows from Corollary 42 that every function $f : X \times Y \to [0, k]$ is $\overline{\mathcal{X} \times \mathcal{Y}}$-measurable, if $|X|$, $|Y|$ are finite.

Definition 43. *Let $(X, \mathcal{X}, \mu)$ and $(Y, \mathcal{Y}, \nu)$ be two fuzzy measure spaces, and $(X \times Y, \overline{\mathcal{X} \times \mathcal{Y}})$ be an extended fuzzy measurable space. We define the fuzzy measures $\overline{m}$ and $\underline{m}$ on $\overline{\mathcal{X} \times \mathcal{Y}}$ induced by μ and ν by*

$$\overline{m}(C) := \sup\{ \oplus_{i \in I} \mu(A_i) \boxdot \nu(B_i) | C = \cup_{i \in I}(A_i \times B_i), \\ A_i \times B_i \in \overline{\mathcal{X} \times \mathcal{Y}}, I : finite\}$$

and

$$\underline{m}(C) := \inf\{ \oplus_{i \in I} \mu(A_i) \boxdot \nu(B_i) | C = \cup_{i \in I}(A_i \times B_i), \\ A_i \times B_i \in \overline{\mathcal{X} \times \mathcal{Y}}, I : finite\}$$

where each $A_i \times B_i$ and $A_j \times B_j$ are disjoint. We call $\overline{m}(C)$ the upper $\oplus$-fuzzy measure induced by μ and ν, and $\underline{m}(C)$ the lower $\oplus$-fuzzy measure induced by μ and ν.

5 Möbius Transform of a Non Additive Measure

In this section we will present the Möbius transform of a non additive measure and a representation of integral with respect to a non additive measure.

5.1 Basic Definition and Properties

We will present the basic definition and properties of the Möbius transform of a non additive measure on a finite set, that is, we assume that a universal set S is a finite set, then we have $\mathcal{S} = 2^S$.

Definition 44. *[60,5] Let μ be a non additive measure on $(S, \mathcal{S})$, then its Möbius transform m is defined as:*

$$m_\mu(A) := \sum_{B \subseteq A} (-1)^{|A|-|B|} \mu(B) \tag{2}$$

for all $A \subset X$.

Note that the function m is not restricted to the $[0, 1]$ interval.

Given a function m that is a Möbius transform, we can reconstruct the original measure as follows:

$$\mu(A) = \sum_{B \subseteq A} m(B)$$

for all $A \subseteq X$.

Example 10. *Let $S := \{x_1, x_2, x_3\}$ and μ be a non additive measure on $(S, \mathcal{S})$. Then we have the Möbius transform m_μ of μ is computed as follows;*

$m_\mu(X) = \mu(X) - \mu(\{x_1, x_2\}) - \mu(\{x_2, x_3\}) - \mu(\{x_3, x_1\}) + \mu(\{x_1\}) + \mu(\{x_2\}) + \mu(\{x_3\})$

$m_\mu(\{x_1, x_2\}) = \mu(\{x_1, x_2\}) - \mu(\{x_1\}) - \mu(\{x_2\})$

$m_\mu(\{x_2, x_3\}) = \mu(\{x_2, x_3\}) - \mu(\{x_2\}) - \mu(\{x_3\})$

$m_\mu(\{x_3, x_1\}) = \mu(\{x_1, x_3\}) - \mu(\{x_1\}) - \mu(\{x_3\})$

Concerning the integral with respect to a non-additive measure, we have the next representation theorem using the Möbius transform.

Theorem 45. *Let m be Möbius transform of μ, we have*

$$(C) \int f d\mu = \sum_{K \subset X} (\inf_{x \in K} f(x)) m(K)$$

It is important for applications that a non additive measure can be identified by means of a linear regression model using the Möbius transform.

5.2 Generalization of Möbius Transform

Three approaches of generalization of Möbius transform on infinite sets and the representation of Choquet integral are known. They are an algebraic way [15,11] , a measure theoretic way[33] and a topological way [42].

First we will introduce the interpreter representation by Murofushi and Sugeno. The interpreter representation theorem is proved in the measure theoretic approach. All proofs are shown in [32,33].

Let $(X, \mathcal{X})$ and $(Y, \mathcal{Y})$ be measurable spaces. A mapping $\mathcal{H} : \mathcal{X} \longrightarrow \mathcal{Y}$ is called an interpreter from $\mathcal{X}$ to $\mathcal{Y}$ if $\mathcal{H}$ satisfies (1) $\mathcal{H}(\emptyset) = \emptyset$ (2) $\mathcal{H}(A) \subset \mathcal{H}(B)$ whenever $A \subset B$. A triplet $(Y, \mathcal{Y}, \mathcal{H})$ is called a frame of $(X, \mathcal{X})$ if $\mathcal{H}$ is an interpreter from $\mathcal{X}$ to $\mathcal{Y}$.

Let μ be a non additive measure on $(X, \mathcal{X})$. A quadruplet $(Y, \mathcal{Y}, m, \mathcal{H})$ is called an interpreter representation of μ if $\mathcal{H}$ is an interpreter from $\mathcal{X}$ to $\mathcal{Y}$, m is a classical measure on $(Y, \mathcal{Y})$ and $\mu = m \circ H$.

For a non negative measurable function f on X we define a function i_f on Y by $i_f(y) := \sup\{r | y \in \mathcal{H}(\{f \geq r\})\}$. We call i_f an interpreter for a measurable function f induced by $\mathcal{H}$.

A semifilter θ in a measurable space $(X, \mathcal{X})$ is a non empty subclass of $\mathcal{X}$ with the properties (a) $\emptyset \notin \theta$ (b) If $A \in \theta$ and $A \subset B \in \mathcal{X}$ then $B \in \theta$. S_X denote the set of all semifilters in $(X, \mathcal{X})$, and define a mapping $H_X : \mathcal{X} \longrightarrow 2^{S_X}$ by $H_X(A) := \{\theta \in S_X | A \in \theta\}$. $\mathcal{S}_X$ denotes the algebra generated by $\{H_X(A) | A \in \mathcal{X}\}$. The triplet $(S_X, \mathcal{S}_X, H_X)$ is called the universal frame of $(X, \mathcal{X})$ for representation.

Theorem 46. (Interpreter representation theorem [32,33]) *Let μ be a non additive measure on $(X, \mathcal{X})$.*

There exists a classical measure m on $\mathcal{S}_X$ such that $(S_X, \mathcal{S}_X, m, H_X)$ is an interpreter representation of μ,and

$$(C) \int f d\mu = \int i_f dm$$

for $f \in \mathcal{L}^{\infty+}$.

Remark. Let μ be a non additive measure on $(X, \mathcal{X})$. We say that $(X, \mathcal{X}, \mu)$ is a fuzzy measure space. Let $(Y, \mathcal{Y}, m, \mathcal{H})$ be an interpreter representation of a fuzzy measure space $(X, \mathcal{X}, \mu)$. The semantics of the interpreter representation are as follows. The element of Y is a feature in respect of the attribute. If $a \in \mathcal{H}(A)$ for $A \in \mathcal{X}$, we say that a is a feature of A or A has the feature a. The measure of A equals the sum of all measures of the feature of A. The interpreter representation is not unique, but all representations are equivalent to the interpreter representation with semifilter (See Appendix in [32]).

Next we present the representation with the Möbius transform by Denneberg. The essences of the proofs are shown in [11].

Let u_A be defined by a $0 - 1$ necessity measure, or a unanimity game for coalition A. $\mathcal{FM}^1_{0u}$ denotes the set of all unanimity games. Next define the tilde operator $\tau_f : \mathcal{FM}^1_{0u} \to R^+$ which assigns $\eta \in \mathcal{FM}^1_{0u}$ to a measurable function $f \in \mathcal{L}^{\infty+}$ by

$$\tau_f(\eta) := (C) \int f d\eta.$$

If $A \in \mathcal{X}$, $\widetilde{A}$ is defined by $\widetilde{A} := \{\eta \in \mathcal{FM}^1_{0u} | \eta(A) = 1\}$. We use the notation $\widetilde{\mathcal{T}} := \{\widetilde{A} | A \in \mathcal{T}\}$ for a class $\mathcal{T} \subset 2^X$. The algebra generated by $\widetilde{\mathcal{X}}$ in $2^{\mathcal{FM}^1_{0u}}$ denoted by $\mathcal{D}_u \subset 2^{\mathcal{FM}^1_{0u}}$. Define the function $\kappa_\eta : \mathcal{FM}^1_{0u} \times \mathcal{X} \to [0, b]$ by

$\kappa_\eta(A) := \eta(A)$ for $\eta \in \mathcal{FM}_{0u}^1$ and $A \in \mathcal{X}$. κ_η is called the zeta function for $\mathcal{X}$. Let ν be a classical measure on $\mathcal{D}_u$. We define the zeta transform μ on $\mathcal{X}$ of ν by

$$\mu(A) := \int \kappa_\eta(A) d\nu(\eta), A \in \mathcal{X}.$$

Theorem 47. (Representation theorem with Möbius transform [11]) *For any non-additive measure μ on $\mathcal{X}$ there exists a unique additive set function ν on $\mathcal{D}_u$ so that*

$$(C)\int f d\mu = \int \tau_f d\nu$$

for $f \in \mathcal{L}^{\infty +}$.

The unique additive set function ν is called the Möbius transform of μ on $\mathcal{D}_u$ and denote it ν^μ. The Möbius transform is not always monotone. It is a signed additive set function on $\mathcal{D}_u$.

5.3 Interpreter and Tilde Operator

In this section, we define a mediator for representations and show the identical points and the differences between the interpreter representation and the representation with a Möbius transform.

A 0-1 fuzzy measure on $\mathcal{X}$ is a $\{0,1\}$−valued fuzzy measure. $\mathcal{FM}_0^1$ denotes the set of all 0-1 fuzzy measures. It is obvious that $\mathcal{FM}_{0u}^1 \subset \mathcal{FM}_0^1$.

Lemma 48. *Let S_X be the class of semifilters of X. There exists a bijection $\varphi : \mathcal{FM}_0^1 \to S_X$.*

We call the bijection φ in the previous lemma a *mediator* for representation.

Theorem 49. *Let $(S_X, \mathcal{S}_X, H_X)$ be the universal frame of $(X, \mathcal{X})$ for representation, and $\varphi : \mathcal{FM}_0^1 \to S_X$ be a mediator for representation.*

1. *$H_X(A) = \varphi(\widetilde{A})$ for $A \in \mathcal{X}$.*
2. *Let $f \in \mathcal{L}^{\infty +}$, i be an interpreter from $\mathcal{X}$ to $\mathcal{S}_X$ and τ_f be a tilde operator from $\mathcal{X}$ to $\widetilde{\mathcal{X}}$. Then we have $i_f \circ \varphi = \tau_f$.*

Let $\mathcal{D}$ be the algebra generated by $\widetilde{\mathcal{X}}$ in $2^{\mathcal{FM}_0^1}$. It is not proved in [11] that there exists an additive set function on $\mathcal{D}$ so that it represents a non additive measure and the Choquet integral. Using the interpreter representation theorem, we can show the existence.

Theorem 50. *For any non-additive measure μ on $\mathcal{X}$ there exists a measure on $\mathcal{D}$ so that*

$$(C)\int f d\mu = \int \widetilde{f} d\nu$$

for every $f \in \mathcal{L}^{\infty +}$.

5.4 Representation on Finite Spaces

In this section, we assume that the universal set X is a finite set. Then the algebra $\mathcal{S}_X$ generated by $\{H_X(A)|A \in \mathcal{X}\}$ is $2^{\mathcal{S}_X}$, and the algebra $\mathcal{D}_u$ generated by $\widetilde{\mathcal{X}}$ is equal to the set $2^{\mathcal{FM}^1_{0u}}$.

First we define another representation, called chain representation.

Definition 51. *Let μ be a non additive measure on $(X, 2^X)$. Since 2^X is finite, the set $\{\mu(A)|A \in 2^X\}$ is a finite subset of real numbers. Therefore we may write*

$$\{\mu(A)|A \in 2^X\} := \{t_i|0 = t_1 < \cdots < t_l = 1\}$$

where l is a positive integer such that $l \leq |2^X|$. Let $\theta_i := \{A \in 2^X|\mu(A) \geq t_i\}$ for $1 \leq i \leq l$. It is obvious from the definition that θ_i is a semifilter. Then using a mediator φ, a non additive measure μ is represented as

$$\mu = \sum_{i=1}^{l} c_i \varphi^{-1}(\theta_i),$$

where c_i is a positive real number. We say that the representation above is a chain representation of a non additive measure μ.

Proposition 52. *[46] A chain representation is one of the special case of the interpreter representation.*

Next we study a belief function of theory of evidence [63]. We say a non additive measure Bel on $(X, 2^X)$ is a belief function if there exists a probability measure m on $(2^X, 2^{2^X})$ such that $Bel(A) = \sum_{B \subset A} m(\{B\})$ for $A \in 2^X$. The probability measure m is called a basic probability assignment (for short b.p.a.).

Proposition 53. *[46] The interpreter representation is a generalization of a representation of b.p.a.*

As an application of the mediator we can show the next well known fact.

Corollary 54. *Let μ be a non additive measure on $(X, \mathcal{X})$. The Möbius transform ν^μ of μ is monotone, that is ν^μ is a probability, if and only if μ is a belief function.*

Example 11. *Let $X := \{a, b, c\}$ and $\mathcal{X} := 2^X$.*

1. *The semifilter $S_X := \{\theta_i\}, (i = 1, 2, \ldots, 15)$ is defined as the Table 1 below. In the Table 1 a number 1 means that a set is an element of the semifilter θ_i and the number 0 means that it is not the element of the semifilter θ_i. Then we have the mapping $H_X(A) := \{\theta \in S_X|A \in \theta\}$ for $A \subset X$ is a vertical line, for example, $H_X(\{a\}) = \{\theta_1, \theta_3, \theta_4, \theta_6, \theta_9\}$.*
   Regarding the Table 1 as a matrix A and using a probability m on $\mathcal{S}_X$. A non additive measure μ is represented as :

$$(\mu(\{a\}), \mu(\{b\}), \ldots, \mu(\{a, c, \}), \mu(\{a, b, c\}))$$
$$= (m(\{\theta_1\}, m(\{\theta_2\}, \ldots, m(\{\theta_{18}\}))A.$$

Table 1. Semifilter

	$\{a\}$	$\{b\}$	$\{c\}$	$\{a,b\}$	$\{b,c\}$	$\{c,a\}$	$\{a,b,c\}$
θ_1	1	1	1	1	1	1	1
θ_2	0	1	1	1	1	1	1
θ_3	1	0	1	1	1	1	1
θ_4	1	1	0	1	1	1	1
θ_5	0	0	1	1	1	1	1
θ_6	1	0	0	1	1	1	1
θ_7	0	1	0	1	1	1	1
θ_8	0	0	1	0	1	1	1
θ_9	1	0	0	1	0	1	1
θ_{10}	0	1	0	1	1	0	1
θ_{11}	0	0	0	1	1	1	1
θ_{12}	0	0	0	0	1	1	1
θ_{13}	0	0	0	1	0	1	1
θ_{14}	0	0	0	1	1	0	1
θ_{15}	0	0	0	1	0	0	1
θ_{16}	0	0	0	0	1	0	1
θ_{17}	0	0	0	0	0	1	1
θ_{18}	0	0	0	0	0	0	1

Table 2. $0-1$ necessity measures

	$\{a\}$	$\{b\}$	$\{c\}$	$\{a,b\}$	$\{b,c\}$	$\{c,a\}$	$\{a,b,c\}$
$u_{\{a\}}$	1	0	0	1	0	1	1
$u_{\{b\}}$	0	1	0	1	1	0	1
$u_{\{c\}}$	0	0	1	0	1	1	1
$u_{\{a,b\}}$	0	0	0	1	0	0	1
$u_{\{b,c\}}$	0	0	0	0	1	0	1
$u_{\{a,c\}}$	0	0	0	0	0	1	1
$u_{\{a,b,c\}}$	0	0	0	0	0	0	1

2. *In the case of* $X := \{a, b, c\}$, *Let* μ *be a* $0-1$ *necessity measure* u_A *for* $A \subset X$ *as represented as in the Table 2:*
   Regarding the Table 2 as a matrix B, *the relation between a non additive measure* μ *and its Möbius transform* ν *is represented as follows:*

$$\begin{aligned}&(\mu(\{a\}), \mu(\{b\}), \dots, \mu(\{a, c, \}), \mu(\{a, b, c\}))\\ &= (\nu(\{a\}), \nu(\{b\}), \dots, \nu(\{a, c, \}), \nu(\{a, b, c\}))B.\end{aligned}$$

   Since $\mathrm{rank}B = 7$, *the Möbius transform is uniquely determined though* μ *is not always monotone.*
3. *Let* μ *be a non additive measure as the Table 3 below.*

Table 3. Fuzzy measure

	$\{a\}$	$\{b\}$	$\{c\}$	$\{a,b\}$	$\{b,c\}$	$\{c,a\}$	$\{a,b,c\}$
μ	0.2	0.3	0.4	0.7	0.7	0.9	1

Define a semifilter $\theta(a)$ *by* $\theta(a) := \{A|\mu(A) \geq a\}$. *The Table 4 shows the semifilters which are used in the chain representation.*

Table 4. Semifilter

	$\{a\}$	$\{b\}$	$\{c\}$	$\{a,b\}$	$\{b,c\}$	$\{c,a\}$	$\{a,b,c\}$
$\theta(0.2)$	1	1	1	1	1	1	1
$\theta(0.3)$	0	1	1	1	1	1	1
$\theta(0.4)$	0	0	1	1	1	1	1
$\theta(0.7)$	0	0	0	1	1	1	1
$\theta(0.9)$	0	0	0	0	0	1	1
$\theta(1)$	0	0	0	0	0	0	1

Regarding the Table 4 as the matrix C, *the coefficients* $c_i, (1 \leq i \leq 6)$ *are determined by the next equation:*

$$(\mu(\{a\}), \mu(\{b\}), \dots, \mu(\{a, c, \}), \mu(\{a, b, c\})) = (c_1, c_2, \dots, c_6)C.$$

As shown in the example above, the interpreter representation cannot determined the probability uniquely, but in [33] Murofushi and Sugeno show that some equivalent relation holds among the representing probability.

Considering the Choquet integral representation with Möbius transform, we have

$$(C)\int f d\mu = \sum_{K \subset X} (\inf_{x \in K} f(x))\nu^{\mu}(u_K) \cdots\cdots (L)$$

where $f \in \mathcal{L}^{\infty+}$ and ν^μ is the Möbius transform of μ. To identify the non additive measure μ, the equation (L) is a simple linear form although ν^μ is not monotone. So it is rather easy to identify the Möbius transform of μ. But it is difficult to interpret the results since ν^μ is signed. Using a mediator φ, we have $\theta = \varphi(u)$ for $\theta \in S_X$ and $u \in \mathcal{FM}^1_{0u}$, and $m \circ H_X(A) = \nu^\mu(\varphi(\widetilde{A}))$, where m is representing classical measure. As shown in [32], the interpreter representation has its own meaning. So it is not difficult to illustrate the meaning of the result. Therefore one can identify the non additive measure easily by the representation with Möbius transform. Using the mediator one can interpret the result easily by the interpreter representation.

6 Concluding Remarks

We have presented a basic definition of a non additive measure and introduced several definitions of continuity of non additive measure and presented the relation among them.

We have introduced the definitions of Sugeno integral, Choquet integral and a generalized fuzzy integral with respect to a non additive measure and we have shown several fundamental properties of them.

We have shown the generalized Möbius transform of a non additive measure and a representation of Choquet integral using the Möbius transform.

The results presented in this Chapter are important for the applications, which will be shown in the rest of this book.

References

1. Aumann, R.J., Shapley, L.S.: Values of Non-atomic Games. Princeton Univ. Press, Princeton (1974)
2. Benvenuti, P., Mesiar, R., Vivona, D.: Monotone set functions-based integrals. In: Pap, E. (ed.) Handbook of Measure Theory, pp. 1329–1379. Elsevier, Amsterdam (2002)
3. Chateauneuf, A.: Modeling attitudes towards uncertainty and risk through the use of Choquet integral. Annals of Operation Research 52, 3–20 (1994)
4. Chateauneuf, A.: Decomposable Measures, Distorted Probabilities and Concave Capacities. Mathematical Social Sciences 31, 19–37 (1996)
5. Chateauneuf, A., Jaffray, J.-Y.: Some characterizations of lower probabilities and other monotone capacities through the use of Möbius inversion. Mathematical Social Sciences 17(3), 263–283 (1989)
6. Choquet, G.: Theory of capacities. Ann. Inst. Fourier, Grenoble. 5, 131–295 (1955)
7. Couceiro, M., Marichal, J.-L.: Characterizations of discrete Sugeno integrals as polynomial functions over distributive lattices. Fuzzy Sets and Systems 161, 694–707 (2010)
8. Dellacherie, C.: Quelques commentaires sur les prolongements de capacités. In: Séminaire de Probabilités 1969/1970, Strasbourg. Lecture Notes in Mathematics, vol. 191, pp. 77–81. Springer (1971)
9. Dempster, A.P.: Upper and lower probabilities induced by a multivalued mapping. Ann. Math. Stat. 38, 325–339 (1967)

10. Denneberg, D.: Non additive measure and Integral. Kluwer Academic Publishers, Dordrecht (1994)
11. Denneberg, D.: Representation of the Choquet integral with the σ-additive Möbius transform. Fuzzy Sets and Systems 92, 139–156 (1997)
12. Dobrakov, I.: On submeasure I. Dissertationes Mathematicae 112, 5–35 (1974)
13. Drewnowski, L.: Topological rings of sets, continuous set functions, integration II. Bull. Acad. Polon. Sci. Ser. Math. Astronom. Phys. 20, 277–286 (1972)
14. Fujishige, S.: Submodular Functions and Optimization, 2nd edn. Elsevier, Amsterdam (2005)
15. Gilboa, I., Schmeidler, D.: Additive representations of non additive measures and the Choquet integral. Ann. Oper. Res. 52, 43–65 (1994)
16. Grabisch, M.: The symmetric Sugeno integral. Fuzzy Sets and Systems 139, 473–490 (2003)
17. Halmos, P.R.: Measure theory. Van Nostrand, Princeton (1950)
18. Klement, E.P., Mesiar, R., Pap, E.: Triangular Norms. Kluwer Academic Publisher, Dordrecht (2000)
19. Kolmogorov, A.N.: Foundations of the theory of probability. Chelsea Publishing Company, New York (1950)
20. König, H.: Measure and integration: an advanced course in basic procedures and applications. Springer, Berlin (1997)
21. Lebesgue, H.: Integrale, longueur, aire. Annali Di Matematica Pura Ed Applicata, 231–359 (1902)
22. Li, J.: Order continuous of monotone set function and convergence of measurable functions sequence. Appl. Math. Comput. 135, 211–218 (2003)
23. Li, J.: A further investigation for Egoroff's theorem with respect to monotone set functions. Kybernetika 39, 753–760 (2003)
24. Li, J.: On Egorofff 's theorems on fuzzy measure spaces. Fuzzy Sets and Systems 135, 367–375 (2003)
25. Li, J., Yasuda, M.: Egoroff 's theorem on monotone non-additive measure spaces. Internat. J. Uncertain. Fuzziness Knowledge-Based Systems 12, 61–68 (2004)
26. Li, J., Yasuda, M.: Lusin's theorem on fuzzy measure spaces. Fuzzy sets and Systems 146, 121–133 (2004)
27. Ling, C.H.: Representation of associative functions. Publ. Math. Debrecen 12, 189–212 (1965)
28. Loväsz, L.: Submodular functions and convexity. In: Bachem, A., Grotschel, M., Korte, B. (eds.) Mathematical Programming: the State of the Art, Bonn 1982, pp. 235–257. Springer, Berlin (1983)
29. Marichal, J.-L.: Weighted lattice polynomials. Discrete Mathematics 309, 814–820 (2009)
30. Menger, K.: Statistical metrics. Proc. Nat. Acad. Sci. 28, 535–537 (1942)
31. Mesiar, R., Vivona, D.: Two-step integral with respect to fuzzy measure. Tatra Mt. Math. Publ. 16, 359–368 (1999)
32. Murofushi, T., Sugeno, M.: An interpretation of fuzzy measures and the Choquet integral as an integral with respect to a fuzzy measure. Fuzzy Sets and Systems 29, 201–227 (1989)
33. Murofushi, T., Sugeno, M.: A theory of Fuzzy Measures: Representations, the Choquet integral, and Null Sets. Journal of Mathematical Analysis and Applications 159(2), 532–549 (1991)
34. Murofushi, T., Sugeno, M.: Fuzzy t-conorm integral with respect to fuzzy measures: Generalization of Sugeno integral and Choquet integral. Fuzzy Sets and Systems 42, 57–71 (1991)

35. Murofushi, T., Sugeno, M.: Hierarchical decomposition of Choquet integral systems. Journal of the Japanese Fuzzy Society 4(4), 749–752 (1992)
36. Murofushi, T., Sugeno, M., Machida, M.: Non-monotonic fuzzy measure and the Choquet integral. Fuzzy Sets and Systems 64(1), 73–86 (1994)
37. Murofushi, T., Narukawa, Y.: A characterization of multi-step discrete Choquet integral. In: 6th Internat. Conference Fuzzy Sets Theory and Its Applications, Abstracts, p. 94 (2002)
38. Murofushi, T., Narukawa, Y.: A characterization of multi-level discrete Choquet integral over a finite set. In: Proc. 7th Workshop on Evaluation of Heart and Mind, pp. 33–36 (2002) (in Japanese)
39. Murofushi, T., Uchino, K., Asahina, S.: Conditions for Egoroff 's theorem in non-additive measure theory. Fuzzy Sets and Systems 146, 35–146 (2004)
40. Murota, K.: Matrices and Matroids for Systems Analysis. Algorithms and Combinatorics, vol. 20. Springer, Berlin (2000)
41. Murota, K.: Discrete Convex Analysis. In: SIAM Monographs on Discrete Mathematics and Applications, vol. 10. Society for Industrial and Applied Mathematics (2003)
42. Narukawa, Y.: A Study of Fuzzy Measure and Choquet Integral, Master Thesis, Tokyo Institute of Technology (1990) (in Japanese)
43. Narukawa, Y., Murofushi, T., Sugeno, M.: Regular fuzzy measure and representation of comonotonically additive functionals. Fuzzy Sets and Systems 112(2), 177–186 (2000)
44. Narukawa, Y., Murofushi, T.: The n-step Choquet integral on finite spaces. In: Proc. 9th Internat. Conference Information Processing and Management of Uncertainty in Knowledge-based Systems, pp. 539–543 (2002)
45. Narukawa, Y., Murofushi, T.: Choquet integral representation and preference. In: Proc. 9th Intern. Conf. Information Processing and Management of Uncertainty (IPMU 2002), Annecy, pp. 747–753 (2002)
46. Narukawa, Y., Murofushi, T.: Representation of Choquet integral; interpreter and Mobius transform. International Journal of Uncertainty, Fuzziness and Knowledge-Based Systems 14, 579–589 (2006)
47. Narukawa, Y., Torra, V.: Twofold integral and Multi-step Choquet integral. Kybernetika 40(1), 39–50 (2004)
48. Narukawa, Y., Torra, V.: Fuzzy measure and probability distributions: distorted probabilities. IEEE Trans. on Fuzzy Systems 13(5), 617–629 (2005)
49. Narukawa, Y., Torra, V.: Multidimensional generalized fuzzy integral. Fuzzy Sets and Systems 160, 802–815 (2009)
50. Narukawa, Y.: Distances defined by Choquet integral. In: IEEE International Conference on Fuzzy Systems, July 24-26, London, England CD-ROM [#1159] (2007)
51. Narukawa, Y., Torra, V.: Continuous OWA operator and its calculation. In: Proc. IFSA-EUSFLAT, Lisbon, Portugal, July 20-24, pp. 1132–1135 (2009) (ISBN:978-989-95079-6-8)
52. von Neumann, J., Morgenstern, O.: Theory of Games and Economic Behavior. Princeton University Press, Princeton (1944)
53. Pap, E.: Null-Additive set functions. Kluwer Academic Publishers, Dordrecht (1995)
54. Preston, M.G., Baratta, P.: An Experimental Study of the Auction-Value of an Uncertain Outcome. American Journal of Psychology 61, 183–193 (1948)
55. Puri, M.L., Ralescu, D.: A possibility measure is not a fuzzy measure. Fuzzy Sets and Systems 7, 311–313 (1982)

56. Quiggin, J.: A Theory of Anticipated Utility. Journal of Economic Behavior and Organization 3, 323–343 (1982)
57. Ralescu, D., Sugeno, M.: Fuzzy integral representation. Fuzzy Sets and Systems 84, 127–133 (1996)
58. Rico, A.: Sugeno integral in a finite Boolean algebra. Fuzzy Sets and Systems 159, 1709–1718 (2008)
59. Roman-Flores, H., Flores-Franulič, A., Chalco-Cano, Y.: A Jensen type inequality for fuzzy integrals. Information Sciences 177, 3192–3201 (2007)
60. Rota, G.-C.: On the Foundations of Combinatorial Theory. I. Theory of Möbius Functions. Z. Wahrscheinlichkeitstheorie 2, 340–368 (1964)
61. Schmeidler, D.: Integral representation without additivity. Proceedings of the American Mathematical Society 97, 253–261 (1986)
62. Schmeidler, D.: Subjective probability and expected utility without additivity. Econometrica 57, 517–587 (1989)
63. Shafer, G.: A Mathematical Theory of Evidence. Princeton University Press, Princeton (1976)
64. Šipoš, J.: Non linear integral. Math. Slovaca 29(3), 257–270 (1979)
65. Sugeno, M.: Theory of fuzzy integrals and its applications, Doctoral Thesis, Tokyo Institute of Technology (1974)
66. Sugeno, M., Murofushi, T.: Pseudo-additive measures and integrals. J. Math. Anal. Appl. 122, 197–222 (1987)
67. Sun, Q.: Property (S) of fuzzy measure and Riesz 's theorem. Fuzzy Sets and Systems 62, 117–119 (1994)
68. Takahashi, M., Murofushi, T.: New conditions for the Egoroff theorem in non-additive measure theory. In: Huynh, V.-N., Nakamori, Y., Lawry, J., Inuiguchi, M. (eds.) Integrated Uncertainty Management and Applications. AISC, vol. 68, pp. 83–89. Springer, Heidelberg (2010)
69. Torra, V., The Weighted, O.W.A.: operator. Int. J. of Intel. Systems 12, 153–166 (1997)
70. Tversky, A., Kahneman, D.: Advances in prospect theory: cumulative representation of uncertainty. J. of Risk and Uncertainty 5, 297–323 (1992)
71. Uchino, K., Murofushi, T.: Relations between mathematical properties of fuzzy measures. In: De Baets, B., Kaynak, O., Bilgiç, T. (eds.) IFSA 2003. LNCS, vol. 2715, pp. 27–30. Springer, Heidelberg (2003)
72. Vitali, G.: Sulla definizione di integrale delle funzioni di una variabile. Annali di Matematica Serie IV, Tomo II, 111–121 (1925)
73. Wang, Z.: The autocontinuity of set function and the fuzzy integral. J. Math. Anal. Appl. 99, 195–218 (1984)
74. Wang, Z., Klir, G.: Fuzzy Measure Theory. Plenum, New York (1992)
75. De Waegenaere, A., Wakker, P.: Nonmonotonic Choquet Integrals. Journal of Mathematical Economics 36, 45–60 (2001)
76. Weber, S.: $\perp$-Decomposable Measures and Integrals for Archimedean t-Conorms $\perp$. J. Math. Anal. Appl. 101, 114–138 (1984)
77. Zadeh, L.: Fuzzy sets as a basis for a theory of possibility. Fuzzy Sets and Systems 1, 3–28 (1978)

Integral Sums and Integrals

Radko Mesiar and Andrea Stupňanová

Slovak University of Technology, Faculty of Civil Engineering, Bratislava, Slovakia
{radko.mesiar,andrea.stupnanova}@stuba.sk

Abstract. We discuss a new approach to integration introduced recently by Even and Lehrer and its relationship to several integrals known from the literature. Decomposition integrals are based on integral sums related to some (possibly constraint) systems of set systems, such as finite chains or finite partitions. A special stress is put on the integrals which are simultaneously decomposition integrals and universal integrals in the sense of Klement et al. Several examples illustrate the presented integrals.

Keywords: Choquet integral, decomposition integral, Sugeno integral, universal integral.

1 Introduction

Integral sums based approach to integration has its roots in ancient Greece. It can be traced already in Eudoxus' (around 370 B.C.) exhaustion principle, which was later successfully applied by several great scientists, including Archimedes and Kepler, among others. Considering the basics of modern mathematics, the most applied integral is surely the one proposed by Riemann in 1854 [23]. Riemann integral is defined on special subsets of $\mathbb{R}^n$. In the case $n = 1$, considered special subsets of $\mathbb{R}$ are (closed) subintervals. Recall that for a non-negative function $f : [a,b] \to [0,\infty[$, Riemann integral $\int_a^b f(x)\,dx$ exists if and only if

$$\sup\left\{\sum_{i=1}^{n} c_i\, l(I_i) |\; n \in \mathbb{N}, (I_i)_{i=1}^{n} \text{ is an interval partition of } [a,b], \sum_{i=1}^{n} c_i\, 1_{I_i} \leq f\right\} =$$
$$= \inf\left\{\sum_{i=1}^{n} c_i\, l(I_i) |\; n \in \mathbb{N}, (I_i)_{i=1}^{n} \text{ is an interval partition of } [a,b], \sum_{i=1}^{n} c_i\, 1_{I_i} \geq f\right\} =$$
$$= \; u \in [0,\infty[, \tag{1}$$

and then $\int_a^b f(x)\,dx = u$. Observe that for an interval I, $l(I)$ denotes its length. Lebesgue [10] has generalized Riemann's approach, considering an arbitrary measurable space $(X,\mathcal{A})$, where $X \neq \emptyset$ is a given universe and $\mathcal{A} \subseteq 2^X$ is a σ-algebra of subsets of X. For a σ - additive measure $m : \mathcal{A} \to [0,\infty]$, and an

V. Torra, Y. Narukawa, and M. Sugeno (eds.), *Non-Additive Measures*,
Studies in Fuzziness and Soft Computing 310,
DOI: 10.1007/978-3-319-03155-2_3,

$\mathcal{A}$ - measurable function $f : X \to [0, \infty]$, the corresponding Lebesgue integral is given by

$$\int_X f dm = \sup \left\{ \sum_{i=1}^{n} c_i\, m(A_i) |\ n \in \mathbb{N}, (A_i)_{i=1}^n \text{ is a measurable partition of } X, \right.$$
$$\left. \sum_{i=1}^{n} c_i\, 1_{A_i} \leq f \right\}. \quad (2)$$

Observe that considering $X = [a, b]$, $\mathcal{A} = \mathcal{B}([a, b])$, and $m = \lambda : \mathcal{A} \to [0, \infty]$ being the classical Lebesgue measure, if the Riemann integral $\int_a^b f(x)\, dx$ exists then the equality $\int_a^b f(x)\, dx = \int_X f\, dm$ holds.

As a special instance of Lebesgue integral recall the case when X is finite, $X = \{x_1, \ldots, x_n\}$. Then, in general, $\mathcal{A} = 2^X$ is considered, and each additive measure $m : \mathcal{A} \to [0, \infty]$ is given by n-tuple of weights $w_i = m(\{x_i\})$. Then

$$\int_X f\, dm = \sum_{i=1}^{n} w_i f(x_i), \quad (3)$$

i.e., Lebesgue integral is just a weighted sum.

Later, several other types of integrals based on special integral sums were introduced. In this chapter, we will discuss these integrals, generalizing the Lebesgue approach. We will deal with an arbitrarily given measurable space $(X, \mathcal{A})$, monotone measures and measurable functions on $(X, \mathcal{A})$.

In the next section, we introduce a framework for discussed integrals, and we recall Choquet integral [3], Sugeno integral [30] and Shilkret integral [28]. In Section 3, we recall recently introduced decomposition integrals of Even and Lehrer [6] and their special instances, including some illustrative examples and a generalization of decomposition integral. Section 4 is devoted to the concept of universal integrals [9] and discussion of integrals which are both universal and (generalized) decomposition integrals. Finally, some concluding remarks are added.

2 Integral Sums Based Integrals

Denote by $\mathcal{S}$ the class of all measurable spaces $(X, \mathcal{A})$. For a fixed measurable space $(X, \mathcal{A})$, we denote by $\mathcal{M}_{(X,\mathcal{A})}$ the set of all monotone measures $m : \mathcal{A} \to [0, \infty]$, $m(\emptyset) = 0$ and $m(A) \leq m(B)$ whenever $A \subseteq B$. The set of all

$\mathcal{A}$ - measurable functions $f : X \to [0, \infty]$ will be denoted as $\mathcal{F}_{(X,\mathcal{A})}$. Moreover, we will consider notation

$$\mathcal{M}^{(1)}_{(X,\mathcal{A})} = \{m \in \mathcal{M}_{(X,\mathcal{A})} |\ m(X) = 1\} \text{ and}$$
$$\mathcal{F}^{(1)}_{(X,\mathcal{A})} = \{f \in \mathcal{F}_{(X,\mathcal{A})} |\ \text{range } f \subseteq [0,1]\}.$$

All integrals considered in this chapter can be seen as mappings

$$\mathbf{J} : \bigcup_{(X,\mathcal{A})\in\mathcal{S}} \left(\mathcal{M}_{(X,\mathcal{A})} \times \mathcal{F}_{(X,\mathcal{A})}\right) \to [0, \infty] \tag{4}$$

$$\left(\text{or } \mathbf{J} : \bigcup_{(X,\mathcal{A})\in\mathcal{S}} \left(\mathcal{M}^{(1)}_{(X,\mathcal{A})} \times \mathcal{F}^{(1)}_{(X,\mathcal{A})}\right) \to [0, \infty] \right)$$

which are either characterized by some axioms, or determined by means of some construction. Mappings given by (4) will be called *general functionals.*

A binary operation $\otimes : [0, \infty]^2 \to [0, \infty]$ ($\otimes : [0,1]^2 \to [0,1]$) will be called a *pseudo-multiplication* whenever it is increasing in both components, 0 is its annihilator (i.e., $0 \otimes c = c \otimes 0 = 0$ for each c), and $\infty \otimes \infty > 0$ ($1 \otimes 1 > 0$). A pseudo-multiplication $\otimes : [0,1]^2 \to [0,1]$ possessing a neutral element $e = 1$ (i.e., $1 \otimes c = c \otimes 1 = c$ for each $c \in [0,1]$) is called a *semicopula* [1,5].

Further, a binary operation $\oplus : [0, \infty]^2 \to [0, \infty]$ is called a *pseudo-addition* whenever it is increasing in each coordinate, it is associative, continuous and 0 is its neutral element [31,9]. Note that due to [18], each pseudo-addition $\oplus$ is also commutative. Due to its associativity, we can extend $\oplus$ to be an n-ary operation on $[0, \infty]$, with notation $\bigoplus_{i=1}^n c_i$.

Basic notions of this chapter are integral summands and integral sums.

Definition 1. *Let $(X, \mathcal{A}) \in \mathcal{S}$, $m \in \mathcal{M}_{(X,\mathcal{A})}$, a pseudo-addition $\oplus : [0, \infty]^2 \to [0, \infty]$ and a pseudo-multiplication $\otimes : [0, \infty]^2 \to [0, \infty]$ be fixed. For a constant $c \in [0, \infty]$, $A \in \mathcal{A}$, the function $b(c, A) : X \to [0, \infty]$*

$$b(c, A)(x) = \begin{cases} c & \text{if } x \in A \\ 0 & \text{elsewhere} \end{cases},$$

is called a basic function, and its m- integral $\otimes$-summand is given by

$$b(c, A)_{m,\otimes} = c \otimes m(A).$$

For a finite system $\mathcal{B} = (b(c_i, A_i))_{i=1}^n$ of basic functions, the corresponding m-integral $(\oplus, \otimes)$-sum is given by

$$S(\mathcal{B}, \oplus, \otimes, m) = \bigoplus_{i=1}^{n} b(c_i, A_i)_{m,\otimes} = \bigoplus_{i=1}^{n} (c_i \otimes m(A_i)).$$

Observe that the Lebesgue integral (2) can be now written as

$$\int_X f dm = \sup\left\{ S(\mathcal{B}, +, \cdot, m) \mid \mathcal{B} = (\, b(c_i, A_i)\,)_{i=1}^n ,\right.$$
$$\left. (A_i)_{i=1}^n \text{ is a partition of } X, \sum_{i=1}^n b(c_i, A_i) \leq f \right\}. \tag{5}$$

The formula (5) can be applied also when m is not σ-additive, and thus it results into a general functional in the sense (4). Note that this functional was introduced and discussed by Yang [35] under the name PAN-integral. For deeper discussion of PAN-integral we recommend [34]. General PAN-integral as introduced in [35] is related to a distributive pair $(\oplus, \otimes)$ of pseudo-addition and pseudo-multiplication, with $\otimes$ possessing a neutral element $e \in]0, \infty]$, associative and continuous on $]0, \infty[^2$ (for deeper discussion of pairs $(\oplus, \otimes)$ linked to PAN-integrals seen [15]), and then

$$\mathbf{PAN}_{(\oplus,\otimes)}(m, f) = \sup\left\{ S(\mathcal{B}, \oplus, \otimes, m) \mid \mathcal{B} = (\, b(c_i, A_i)\,)_{i=1}^n ,\right.$$
$$\left. (A_i)_{i=1}^n \text{ is a partition of } X, \bigoplus_{i=1}^n b(c_i, A_i) \leq f \right\}. \tag{6}$$

Note that when considering integral sums based on measurable partitions as in (6), one can obtain the same integral for different monotone measures. Take, for example, $X = \{x_1, x_2\}$, $m_1(\{x_1\}) = m_2(\{x_1\}) = 2$, $m_1(\{x_2\}) = m_2(\{x_2\}) = 3$, $m_1(X) = 4$, $m_2(X) = 5$. Then for any $f \in \mathcal{F}_{(X,2^X)}$,

$$\mathbf{PAN}_{(+,\cdot)}(m_1, f) = \mathbf{PAN}_{(+,\cdot)}(m_2, f) = 2f(x_1) + 3f(x_2),$$

though $m_1 \neq m_2$.

Alternative approaches to integration are related to integral sums based on measurable chains. Paraphrasing (5), considering $(A_i)_{i=1}^n$ to be a chain, the famous Choquet integral [3] is recovered. Note that this integral is given by

$$\mathbf{Ch}(m, f) = \int_0^\infty m(f \geq t) dt, \tag{7}$$

but also

$$\mathbf{Ch}(m, f) = \sup\left\{ S(\mathcal{B}, +, \cdot, m) \mid \mathcal{B} = (\, b(c_i, A_i)\,)_{i=1}^n ,\right.$$
$$\left. (A_i)_{i=1}^n \text{ is a chain }, \sum_{i=1}^n b(c_i, A_i) \leq f \right\}. \tag{8}$$

Similarly, we can restrict to singleton systems $\mathcal{B}$ only, and then Shilkret integral [28] is recovered,

$$\mathbf{Sh}(m, f) = \sup\left\{c \cdot m(A) |\ b(c, A) \leq f\right\}. \tag{9}$$

Formulae (5), (8), (9) deals with $\oplus = +$ and $\otimes = \cdot$ (i.e., with the standard arithmetical operations), while the formula (6) with PAN-operations $(\oplus, \otimes)$. Considering the pair $(\vee, \wedge)$ (i.e., lattice operations of join (maximum) and meet (minimum) on $[0, \infty]$), one obtain in all cases the Sugeno integral [30],

$$\mathbf{Su}(m, f) = \sup\left\{c \wedge m(A) |\ b(c, A) \leq f\right\} = \sup\left\{t \wedge m(f \geq t) |\ t \in [0, \infty]\right\}. \tag{10}$$

In the next example we exemplify some of the above introduced integrals.

Example 1. Let $X = [0, 1]$, $\mathcal{A} = \mathcal{B}([0, 1])$ and $\lambda : \mathcal{A} \to [0, \infty]$ be the standard Lebesgue measure. Then, for $f : [0, 1] \to [0, \infty]$ given by $f(x) = x$,

$$\int_0^1 f(x)\, dx = \int_X f\, d\lambda = \mathbf{PAN}_{(+,\cdot)}(\lambda, f) = \mathbf{Ch}(\lambda, f) = \mathbf{Su}(\lambda, f) = \frac{1}{2}$$

$$\text{and} \quad \mathbf{Sh}(\lambda, f) = \frac{1}{4}.$$

Considering $m = \lambda^2 \in \mathcal{M}_{(X,\mathcal{A})}$, observe that m is not σ-additive and thus the Lebesgue integral cannot be applied. For the applicable integrals we obtain:

$$\mathbf{PAN}_{(+,\cdot)}(\lambda^2, f) = \frac{2\sqrt{3} - 3}{3} = 0.155,$$

$$\mathbf{Ch}(\lambda^2, f) = \int_0^1 (1 - t)^2\, dt = \frac{1}{3} = 0.333,$$

$$\mathbf{Sh}(\lambda^2, f) = \sup\left\{t(1 - t)^2 |\, t \in [0, 1]\right\} = \frac{4}{27} = 0.148,$$

$$\mathbf{Su}(\lambda^2, f) = \sup\left\{t \wedge (1 - t)^2 |\, t \in [0, 1]\right\} = \frac{3 - \sqrt{5}}{2} = 0.386.$$

3 Decomposition Integrals

The idea of decomposition integrals connecting several types of integrals is due to Even and Lehrer [6], and it is related to the standard arithmetical operations $+$ and $\cdot$. This is, for example, the case of Choquet, Shilkret and $PAN_{(+,\cdot)}$ integrals. Before the formal introduction of decomposition integrals, consider first the next optimization example.

Example 2. Consider a group of workers $X = \{x_1, x_2, x_3\}$ with daily performance $m(X) = 8$. For smaller groups of workers, the next daily performances are known: $m(\{x_1\}) = 2, m(\{x_2\}) = 3, m(\{x_3\}) = m(\{x_1, x_3\}) = 4$, $m(\{x_1, x_2\}) = 7$ and $m(\{x_2, x_3\}) = 5$.

The availability of single workers in working days is $f(x_1) = 5$, $f(x_2) = 4$ and $f(x_3) = 3$ (i.e., worker x_1 can work maximally 5 days altogether, though possibly in different groups of workers). How to organize the working plan to attain the maximal global performance, being constraint by one of the following "working laws"?

Working Laws

(i) only one group of workers can work for a fixed time period;
(ii) several disjoint groups of workers can work, each for its fixed time period;
(iii) one group of workers starts to work, a worker after stopping his work cannot start again;
(iv) there are no constraints.

Optimal Performances

(i) we have to compute $\max\{c{\cdot}m(A) | \, b(c, A) \leq f\}$, i.e., the optimal performance is just the Shilkret integral

$$\mathbf{Sh}(m, f) = f(x_2) \cdot m(\{x_1, x_2\}) = 4 \cdot 7 = 28;$$

(ii) we have to compute
$\max\left\{ \sum_{i=1}^{n} c_i m(A_i) \middle| \, \sum_{i=1}^{n} b(c_i, A_i) \leq f, (A_i)_{i=1}^{n} \text{ is a partition of } X \right\}$, i.e., the optimal performance is the PAN-integral

$$\mathbf{PAN}_{(+,\cdot)}(m, f) = f(x_2) \cdot m(\{x_1, x_2\}) + f(x_3) \cdot m(\{x_3\}) = 4 \cdot 7 + 3 \cdot 4 = 40;$$

(iii) we have to compute
$\max\left\{ \sum_{i=1}^{n} c_i m(A_i) \middle| \, \sum_{i=1}^{n} b(c_i, A_i) \leq f, (A_i)_{i=1}^{n} \text{ is a chain in } 2^X \right\}$, i.e., the optimal performance is the Choquet integral

$$\begin{aligned}\mathbf{Ch}(m, f) = f(x_3) \cdot m(X) + (f(x_2) - f(x_3)) \cdot m(\{x_1, x_2\}) + \\ + (f(x_1) - f(x_2)) \cdot m(\{x_1\}) = 3 \cdot 8 + 1 \cdot 7 + 1 \cdot 2 = 33;\end{aligned}$$

(iv) the optimal performance is

$$\begin{aligned}f(x_2) \cdot m(\{x_1, x_2\}) + f(x_3) \cdot m(\{x_3\}) + (f(x_1) - f(x_2)) \cdot m(\{x_1\}) = \\ = 4 \cdot 7 + 3 \cdot 4 + 1 \cdot 2 = 42.\end{aligned}$$

Note that the case (iv) is the above example corresponds to the *concave integral* recently introduced by Lehrer [11,12],

$$\mathbf{L}(m,f)=\sup\left\{\sum_{i=1}^{n}c_i\,m(A_i)\middle|\; n\in\mathbb{N},\ \sum_{i=1}^{n}b(c_i,A_i)\le f\right\}. \tag{11}$$

It is not difficult to see that $\mathbf{Sh}\le\mathbf{Ch}\le\mathbf{L}$ and $\mathbf{Sh}\le\mathbf{PAN}_{(+,\cdot)}\le\mathbf{L}$, independently of a fixed measurable space $(X,\mathcal{A})$, monotone measure $m\in\mathcal{M}_{(X,\mathcal{A})}$ and function $f\in\mathcal{F}_{(X,\mathcal{A})}$. In Example 2, it holds

$$\mathbf{Sh}(m,f)=28<\mathbf{Ch}(m,f)=33<\mathbf{PAN}_{(+,\cdot)}(m,f)=40<\mathbf{L}(m,f)=42.$$

As we can see from formulae (8), (9), (11) and (6) for $(\oplus,\odot)=(+,\cdot)$, all these introduced integrals are based on (sub -)decomposition of the considered function $f\in\mathcal{F}_{(X,\mathcal{A})}$, into acceptable basic functions, $\sum_{i=1}^{n}b(c_i,A_i)\le f$, where the system $(A_i)_{i=1}^{n}$ satisfies some constraints (it is a chain in (8), a singleton in (9), there is no constraint in (11), and it is partition in (6)). This observation was a motivation for Even and Lehrer [6] to introduce decomposition systems and decomposition integrals.

Definition 2. *For a fixed $(X,\mathcal{A})\in\mathcal{S}$, a non-empty system $\mathcal{H}$ of collections from $\mathcal{A}$ (i.e., finite non-empty subsets of $\mathcal{A}$) is called a decomposition system. Moreover, the mapping $\mathbf{I}_{\mathcal{H}}:\mathcal{M}_{(X,\mathcal{A})}\times\mathcal{F}_{(X,\mathcal{A})}\to[0,\infty]$ given by*

$$\mathbf{I}_{\mathcal{H}}(m,f)=\sup\left\{\sum_{i=1}^{n}c_i\,m(A_i)\middle|\;(A_i)_{i=1}^{n}\in\mathcal{H},\ \sum_{i=1}^{n}b(c_i,A_i)\le f\right\} \tag{12}$$

is called a decomposition integral.

Considering a setting of constraints (e.g., only finite chains are considered) such that for each measurable space $(X,\mathcal{A})\in\mathcal{S}$ it determines a unique decomposition system $\mathcal{H}$, we can define by means of (12) a general functional $\mathbf{J}_{\mathcal{H}}$ which will be called *a general decomposition integral* (we have slightly abused the notation, as $\mathcal{H}$ in $\mathbf{J}_{\mathcal{H}}$ means the considered setting of constraints, while $\mathcal{H}$ in $\mathbf{I}_{\mathcal{H}}$ means the considered decomposition system on $(X,\mathcal{A})$.)

Recall that from the integrals we have discussed so far, general decomposition integrals include:

Shilkret integral (9) constraint to singletons, i.e., for any $(X,\mathcal{A})\in\mathcal{S}$,

$$\mathcal{H}_{Sh}=\{\{A\}|\,A\in\mathcal{A}\setminus\{\emptyset\}\};$$

Choquet integral (8) constraint to finite chains, i.e., for any $(X,\mathcal{A})\in\mathcal{S}$,

$$\mathcal{H}_{Ch}=\{\mathcal{C}|\,\mathcal{C}\text{ is a finite chain in }\mathcal{A}\};$$

$\mathbf{PAN}_{(+,\cdot)}$ integral (6) constraint to finite partitions, i.e., for any $(X,\mathcal{A})\in\mathcal{S}$,

$$\mathcal{H}_{PAN}=\{\mathcal{P}|\,\mathcal{P}\text{ is a finite partition of }(X,\mathcal{A})\};$$

concave integral of Lehrer (11) no constraint, i.e., for any $(X, \mathcal{A}) \in \mathcal{S}$,

$$\mathcal{H}_L = \{\mathcal{D} | \mathcal{D} \text{ is a finite non-empty subset of } \mathcal{A} \setminus \{\emptyset\}\}.$$

The next example brings a decomposition integral which is not general, i.e., it is defined only on a fixed measurable space $(X, \mathcal{A}) \in \mathcal{S}$.

Example 3. Continuing in Example 2, suppose the next "working law":

(v) the workers x_1 and x_2 cannot work together.

Then the corresponding decomposition system $\mathcal{H}$ is given by
$\mathcal{H} = \{\mathcal{E} | \emptyset \neq \mathcal{E} \subset 2^X \setminus \{\emptyset\}$, there is no $A \in \mathcal{E}$ such that $\{x_1, x_2\} \subseteq A\}$, and the best performance is attained on singletons, i.e., when each worker works alone,

$$\begin{aligned}\mathbf{I}_{\mathcal{H}}(m, f) =& f(x_1) \cdot m(\{x_1\}) + f(x_2) \cdot m(\{x_2\}) + f(x_3) \cdot m(\{x_3\}) = \\ =& 5 \cdot 2 + 4 \cdot 3 + 3 \cdot 4 = 34.\end{aligned}$$

It is obvious that more restrictive constraints when introducing general decomposition integrals yield to smaller integral output, as already exemplified on the relationships between Shilkret, $\text{PAN}_{(+,\cdot)}$, Choquet and concave integrals.

On the other hand, two different decomposition systems $\mathcal{H}_1$ and $\mathcal{H}_2$ can result into a unique decomposition integral $\mathbf{I}_{\mathcal{H}_1} = \mathbf{I}_{\mathcal{H}_2}$ (general decomposition integral $\mathbf{J}_{\mathcal{H}_1} = \mathbf{J}_{\mathcal{H}_2}$). As a typical example consider (supposing $\text{card}\, X > 1$)

$$\begin{aligned}\mathcal{H}_1 &= \{\mathcal{C} | \mathcal{C} \text{ is a chain in } \mathcal{A} \text{ of length } 2\} \quad \text{and} \\ \mathcal{H}_2 &= \{\mathcal{C} | \mathcal{C} \text{ is a chain in } \mathcal{A} \text{ of length at most } 2\}.\end{aligned}$$

Then

$$\mathbf{J}_{\mathcal{H}_1}(m, f) = \mathbf{J}_{\mathcal{H}_2}(m, f) = \sup\{c \cdot m(f \geq c) + d \cdot m(f \geq c + d) | \, c, d \in [0, \infty]\}.$$

For a deeper discussion of properties of decomposition integrals we recommend [6,16,32].

We introduce two graded families of decomposition systems:

$$\begin{aligned}\mathcal{H}_{(n)} =& \{\mathcal{C} | \mathcal{C} \text{ is a chain in } \mathcal{A} \text{ with length at most } n\}; \\ \mathcal{H}^{(n)} =& \{\mathcal{P} | \mathcal{P} \text{ is a partition of } (X, \mathcal{A}) \text{ with at most } n \text{ members}\}.\end{aligned}$$

Evidently $\mathcal{H}_{(1)} = \mathcal{H}^{(1)} = \mathcal{H}_{Sh}$ is related to the Shilkret integral. Moreover, it holds

$$\mathcal{H}_{(1)} \subsetneq \mathcal{H}_{(2)} \subsetneq \cdots \subsetneq \mathcal{H}_{(n)} \subsetneq \cdots \subsetneq \mathcal{H}_{Ch},$$

$\mathcal{H}_{Ch} = \bigcup_{n=1}^{\infty} \mathcal{H}_{(n)}$, is related to the Choquet integral, and

$$\mathcal{H}^{(1)} \subsetneq \mathcal{H}^{(2)} \subsetneq \cdots \subsetneq \mathcal{H}^{(n)} \subsetneq \cdots \subsetneq \mathcal{H}_{PAN},$$

$\mathcal{H}_{PAN} = \bigcup_{n=1}^{\infty} \mathcal{H}^{(n)}$, is related to the PAN$_{(+,\cdot)}$ integral. Moreover, in [32] we have

introduced a new general decomposition system

$$\mathcal{H}^* = \Big\{\mathcal{D} \mid \emptyset \neq \mathcal{D} \subseteq \mathcal{A} \setminus \{\emptyset\}, \mathcal{D} \text{ is finite, for each } A, B \in \mathcal{D}, \text{ either } A \cap B = \emptyset \text{ or } (A, B) \text{ is a chain}\Big\}.$$

Then, on any $(X, \mathcal{A}) \in \mathcal{S}$, $\mathcal{H}_{PAN} \subsetneq \mathcal{H}^* \subsetneq \mathcal{H}_L$ and $\mathcal{H}_{Ch} \subsetneq \mathcal{H}^* \subsetneq \mathcal{H}_L$. For the corresponding general decomposition integrals, their relationships is visualized on the Hasse diagram depicted in Figure 1.

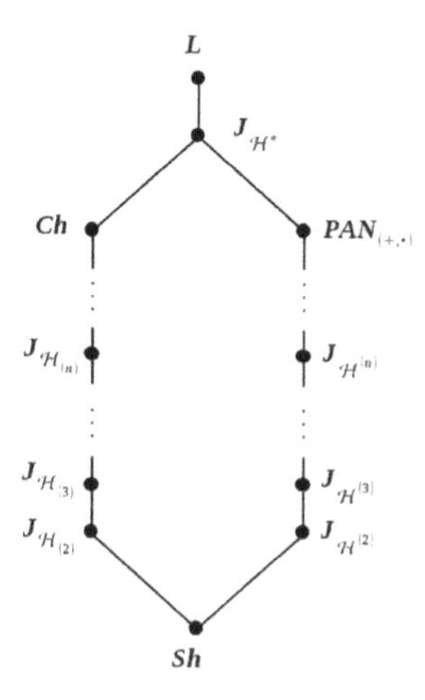

Fig. 1. Hasse diagram relating introduced general decomposition integrals

Remark 1

(1) Decomposition integrals can be seen as an approach to integration based on the lower integral sums (sub-decomposition), expressed in the constraint $\sum_{i=1}^{n} b(c_i, A_i) \leq f$. It is possible to consider a dual approach based on the upper integral sums (super-decomposition), requiring $\sum_{i=1}^{n} b(c_i, A_i) \geq f$. Then the corresponding decomposition integral $\mathbf{I}^{\mathcal{H}} : \mathcal{M}_{(X,\mathcal{A})} \times \mathcal{F}_{(X,\mathcal{A})} \to [0, \infty]$ can be defined as

$$\mathbf{I}^{\mathcal{H}}(m, f) = \inf \left\{ \sum_{i=1}^{n} c_i\, m(A_i) \big| \, (A_i)_{i=1}^{n} \in \mathcal{H}, \ \sum_{i=1}^{n} b(c_i, A_i) \geq f \right\}. \tag{13}$$

(2) Decomposition integrals deal with the classical arithmetics, i.e., with the operations pair $(+, \cdot)$. It is immediate that this approach can be extended to

any pair $(\oplus, \otimes)$ of a pseudo-addition and pseudo-multiplication introducing, for example, $(\oplus, \otimes)$-decomposition integral
$\mathbf{I}_{\mathcal{H}}^{(\oplus,\otimes)} : \mathcal{M}_{(X,\mathcal{A})} \times \mathcal{F}_{(X,\mathcal{A})} \to [0, \infty]$ defined as

$$\mathbf{I}_{\mathcal{H}}^{(\oplus,\otimes)}(m, f) = \sup \left\{ \bigoplus_{i=1}^{n} (c_i \otimes m(A_i)) \middle| (A_i)_{i=1}^{n} \in \mathcal{H}, \bigoplus_{i=1}^{n} b(c_i, A_i) \leq f \right\}. \tag{14}$$

Observe that $\mathbf{I}_{\mathcal{H}}^{(\vee,\wedge)} = \mathbf{Su}$ is the Sugeno integral on $\mathcal{M}_{(X,\mathcal{A})} \times \mathcal{F}_{(X,\mathcal{A})}$ whenever for each non-empty $A \in \mathcal{A}$ there is a collection $\mathcal{D} \in \mathcal{H}$ such that $A \in \mathcal{D}$. For more details and discussion about pseudo-decomposition integrals see, for example, [14].

Note that, in general, there are only few properties valid for each $(\oplus, \otimes)$-decomposition integral. Obviously, considering a general decomposition system $\mathcal{H}$, $\mathbf{I}_{\mathcal{H}}^{(\oplus,\otimes)}$ is a general functional which is increasing in each coordinate. If the pseudo-multiplication $\otimes$ is associative and the pseudo-addition $\oplus$ is left-distributive over $\otimes$, i.e.,

$$a \otimes (b \oplus c) = (a \otimes b) \oplus (a \otimes c),$$

then $\mathbf{I}_{\mathcal{H}}^{(\oplus,\otimes)}$ is $\otimes$-homogeneous functional,

$$\mathbf{I}_{\mathcal{H}}^{(\oplus,\otimes)}(m, c \otimes f) = c \otimes \mathbf{I}_{\mathcal{H}}^{(\oplus,\otimes)}(m, f)$$

for each $c \in]0, \infty[$, $m \in \mathcal{M}_{(X,\mathcal{A})}$ and $f \in \mathcal{F}_{(X,\mathcal{A})}$.

Therefore, decomposition integrals are (positively) homogeneous.

4 Universal Integrals

Decomposition integrals can be seen as a constructive approach to integrals, though in several cases we do not know an effective algorithm how to evaluate them exactly. Moreover, they can have some undesirable properties. For example, it might happen that $\mathbf{I}_{\mathcal{H}}(m, 1_A) > m(A)$. Indeed, considering Example 1, $\mathbf{PAN}_{(+,\cdot)}(m, 1_{\{x_2,x_3\}}) = \mathbf{L}(m, 1_{\{x_2,x_3\}}) = 7$ while $m(\{x_2, x_3\}) = 5$. In 2010, we have proposed in [9] an axiomatic approach to integration, giving a framework to functionals deserving to be called integrals.

Definition 3. *A general functional* $\mathbf{J} : \bigcup_{(X,\mathcal{A}) \in \mathcal{S}} \left(\mathcal{M}_{(X,\mathcal{A})} \times \mathcal{F}_{(X,\mathcal{A})}\right) \to [0, \infty]$ *is called a universal integral whenever the next axioms are satisfied:*

(U1) $\mathbf{J}$ *is increasing in both components;*
(U2) $\mathbf{J}(m, b(c, A))$ *depends on* c *and* $m(A)$ *only, independently of the measurable space* $(X, \mathcal{A}) \in \mathcal{S}$, $m \in \mathcal{M}_{(X,\mathcal{A})}$ *and* $c \in [0, \infty]$, *and there is a constant* $e \in]0, \infty]$ *such that for all* $(X, \mathcal{A}) \in \mathcal{S}$,

$$\mathbf{J}(m, b(e, A)) = m(A)$$

for all $m \in \mathcal{M}_{(X,\mathcal{A})}$, $A \in \mathcal{A}$ *and*

$$\mathbf{J}(m, b(c, X)) = c$$

for all $c \in [0, \infty]$ *and all* $m \in \mathcal{M}_{(X,\mathcal{A})}$ *such that* $m(X) = e$.

(U3) $\mathbf{J}(m_1, f_1) = \mathbf{J}(m_2, f_2)$ *for all pairs* $(m_1, f_1) \in \mathcal{M}_{(X_1,\mathcal{A}_1)} \times \mathcal{F}_{(X_1,\mathcal{A}_1)}$, $(m_2, f_2) \in \mathcal{M}_{(X_2,\mathcal{A}_2)} \times \mathcal{F}_{(X_2,\mathcal{A}_2)}$ *with the same survival function* $h_{m_1,f_1} = h_{m_2,f_2}$, *where* $h_{m,f} :]0, \infty] \to [0, \infty]$ *is given by*

$$h_{m,f}(t) = m(f \geq t).$$

Note that pairs (m_i, f_i) with a common survival function were called *integral equivalent* in [9]. In the probability theory, such pairs (probability measure and a non-negative random variable) are characterized by a common distribution function on $]0, \infty]$, and the corresponding survival function is its complement to 1.

The axiom (U2) ensures the existence of a pseudo-multiplication $\otimes : [0, \infty] \to [0, \infty]$ with a neutral element $e \in]0, \infty]$, such that $\mathbf{J}(m, b(c, A)) = c \otimes m(A)$. The axiom (U1) allows to introduce, for any pseudo-multiplication $\otimes$ on $[0, \infty]$ with a neutral element $e \in]0, \infty]$, the smallest universal integral $\mathbf{J}_\otimes$ satisfying (U2) with $\otimes$,

$$\mathbf{J}_\otimes(m, f) = \sup \{t \otimes m(f \geq t) | \, t \in]0, \infty]\}. \tag{15}$$

Evidently $\mathbf{J}_\cdot = \mathbf{Sh}$ is the Shilkret integral and $\mathbf{J}_\wedge = \mathbf{Su}$ is the Sugeno integral.

Similarly, one can introduce the greatest universal integral $\mathbf{J}^\otimes$ satisfying (U2) with $\otimes$,

$$\begin{aligned} \mathbf{J}^\otimes(m, f) &= \inf \{c \otimes m(A) | \, b(c, A) \geq f\} = \\ &= \operatorname{essup}_m(f) \otimes \sup \{m(f \geq t) | \, t \in]0, \infty]\}, \end{aligned} \tag{16}$$

where $\operatorname{essup}_m(f) = \sup \{t \in [0, \infty] \, | m(f \geq t) > 0\}$ and $\sup \{m(f \geq t) | \, t \in]0, \infty]\} = \lim_{t \to 0^+} m(f \geq t)$.

For more details and some construction methods for universal integrals we recommend [9].

The increase of interest in new, in general non-additive integrals can be traced back to the introduction of fuzzy sets [36] and attempts to introduce expectations for fuzzy sets [37,30]. Nowadays these integrals are substantially exploited in several multicriteria decision problems dealing with graded scales for considered criteria. The most applied scale is the unit interval $[0, 1]$, and for this purpose we have introduced the concept of universal integrals on $[0, 1]$, too. For more details see [9].

Definition 4. *A mapping* $\mathbf{J} : \bigcup_{(X,\mathcal{A}) \in \mathcal{S}} \left(\mathcal{M}^{(1)}_{(X,\mathcal{A})} \times \mathcal{F}^{(1)}_{(X,\mathcal{A})}\right) \to [0, 1]$ *is called a universal integral on* $[0, 1]$ *whenever the next axioms are satisfied:*

(U1) *(i.e.,* $\mathbf{J}$ *is increasing in both components);*
(U2') *there is a semicopula* $T : [0,1]^2 \to [0,1]$ *so that for any* $(X, \mathcal{A}) \in \mathcal{S}$, $\mathbf{J}(m, b(c, A)) = T(c, m(A))$ *for each* $m \in \mathcal{M}^{(1)}_{(X,\mathcal{A})}$, $A \in \mathcal{A}$ *and* $c \in [0,1]$;
(U3) *(i.e.,* $\mathbf{J}(m_1, f_1) = \mathbf{J}(m_2, f_2)$ *whenever* $h_{m_1,f_1} = h_{m_2,f_2}$*).*

Also in this case the smallest universal integral $\mathbf{J}_T$ and the greatest universal integral $\mathbf{J}^T$ satisfying (U2') with a given semicopula T can be introduced, applying the formulas (15) and (16), respectively. In a special case, considering a supermodular semicopula $C : [0,1]^2 \to [0,1]$, i.e., a copula [21], observe first that there is a one-to-one correspondence between copulas and probabilities on Borel subsets of $[0,1]^2$ with uniformly distributed margins, namely $P_C([0,u] \times [0,v]) = C(u,v)$, $(u,v) \in [0,1]^2$. This fact allows to introduce a new class of universal integrals on $[0,1]$, see [8,9].

Proposition 1. *Let* $C : [0,1]^2 \to [0,1]$ *be a fixed copula. Then the mapping* $\mathbf{J}_{(C)} : \bigcup_{(X,\mathcal{A}) \in \mathcal{S}} \left(\mathcal{M}^{(1)}_{(X,\mathcal{A})} \times \mathcal{F}^{(1)}_{(X,\mathcal{A})}\right) \to [0,1]$ *given by*

$$\mathbf{J}_{(C)}(m,f) = P_C\left(\{(x,y) \in]0,1]^2 | \, y \leq h_{m,f}(x)\}\right) \tag{17}$$

is a universal integral on $[0,1]$.

It is not difficult to check that for the strongest copula M, $M(u,v) = \min\{u,v\}$,

$$\mathbf{J}_{(M)} = \mathbf{Su}$$

is the Sugeno integral in its original form acting on $[0,1]$, as introduced in [30]. Considering the product copula Π, $\Pi(u,v) = uv$, one gets

$$\mathbf{J}_{(\Pi)} = \mathbf{Ch},$$

i.e., the Choquet integral is recovered (restricted to $\bigcup_{(X,\mathcal{A}) \in \mathcal{S}} \left(\mathcal{M}^{(1)}_{(X,\mathcal{A})} \times \mathcal{F}^{(1)}_{(X,\mathcal{A})}\right)$).

Observe that for two copulas C_1, C_2, also their convex combination $C = \alpha C_1 + (1-\alpha) C_2$, $\alpha \in [0,1]$, is a copula, and then

$$\mathbf{J}_{(C)} = \alpha \mathbf{J}_{(C_1)} + (1-\alpha)\mathbf{J}_{(C_2)}$$

is the corresponding convex combination of copula - based universal integrals on $[0,1]$. This observation allows to apply convex combinations of M and Π to generate a parametric class of integrals $(\lambda \mathbf{Su} + (1-\lambda)\mathbf{Ch} | \, \lambda \in [0,1])$ which can be of interest for fitting purposes when modeling the real world problems.

In [16], we have studied general integrals which are simultaneously decomposition integrals and universal integrals.

Proposition 2. *A general decomposition integral* $\mathbf{J}_{\mathcal{H}} : \bigcup_{(X,\mathcal{A}) \in \mathcal{S}} \left(\mathcal{M}_{(X,\mathcal{A})} \times \mathcal{F}_{(X,\mathcal{A})}\right) \to [0,\infty]$ *is a universal integral if and only if*

$$\mathcal{H} \in \{\mathcal{H}_{(n)} | \, n \in \mathbb{N}\} \cup \{\mathcal{H}_{Ch}\}.$$

Obviously, all integrals characterized by Proposition 2 are linked to the classical multiplication.

Example 4. Continuing in Example 1, for integrals characterized in Proposition 2, it holds:

$$\mathbf{J}_{\mathcal{H}_{(1)}}(\lambda, f) = \mathbf{Sh}(\lambda, f) = \frac{1}{4} \;\; (\text{ attained for } b\left(\frac{1}{2}, 1_{[\frac{1}{2},1]}\right)),$$

$$\mathbf{J}_{\mathcal{H}_{(2)}}(\lambda, f) = \frac{1}{3} \;\; (\text{ attained for } b\left(\frac{1}{3}, 1_{[\frac{1}{3},1]}\right) \text{ and } b\left(\frac{1}{3}, 1_{[\frac{2}{3},1]}\right)),$$

$$\vdots$$

$$\mathbf{J}_{\mathcal{H}_{(n)}}(\lambda, f) = \frac{n}{2(n+1)} \;\; (\text{ attained for } b\left(\frac{1}{n+1}, 1_{[\frac{1}{n+1},1]}\right),$$

$$b\left(\frac{1}{n+1}, 1_{[\frac{2}{n+1},1]}\right), \;\dots\;, b\left(\frac{1}{n+1}, 1_{[\frac{n}{n+1},1]}\right)),$$

with limit member

$$\mathbf{Ch}(\lambda, f) = \lim_{n\to\infty} \mathbf{J}_{\mathcal{H}_{(n)}}(\lambda, f) = \frac{1}{2}.$$

In this example we can explicitely see the interpolative character of the class $\left(\mathbf{J}_{\mathcal{H}_{(n)}}\right)_{n=1}^{\infty}$ of the integrals introduced in Proposition 2, varying from the Shilkret integral $\mathbf{Sh} = \mathbf{J}_{\mathcal{H}_{(1)}}$ to the Choquet integral $\mathbf{Ch} = \lim_{n\to\infty} \mathbf{J}_{\mathcal{H}_{(n)}}$. For visualization see Figure 2.

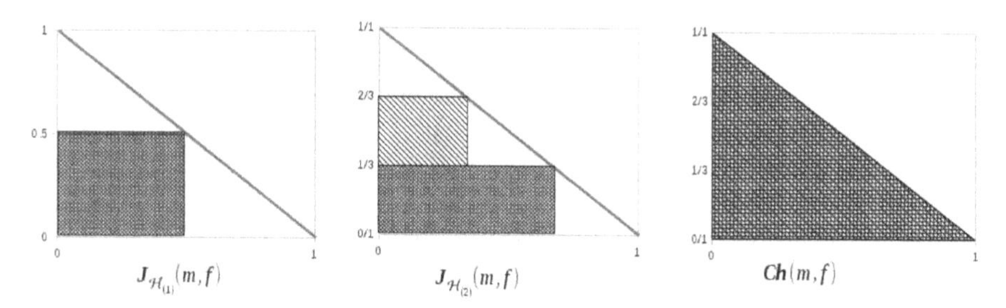

Fig. 2. Shaded areas correspond to integrals $\mathbf{J}_{\mathcal{H}_{(1)}}$, $\mathbf{J}_{\mathcal{H}_{(2)}}$ and **Ch**

Proposition 2 can be rewritten for universal integrals on $[0, 1]$ with no substantial changes. Moreover, following the results of Proposition 2, we can define new classes of universal integrals on $[0, 1]$ based on copulas and decomposition systems, generalizing the equality

$$\mathbf{J}_{\mathcal{H}_{(n)}}(m, f) = \sup\left\{\sum_{i=1}^{n} c_i\, m(f \geq c_1 + \cdots + c_i) \,\middle|\, c_1, \dots, c_n \in [0, \infty]\right\}. \tag{18}$$

Note that if (18) aims to describe the universal integral on $[0, 1]$, the constraint $\sum_{i=1}^{n} c_i \leq 1$ should be considered.

Definition 5. *Let $C : [0,1]^2 \to [0,1]$ be a fixed copula. Then, for any $n \in \mathbb{N}$, the mapping $\mathbf{J}_{(C,n)} : \bigcup_{(X,\mathcal{A})\in\mathcal{S}} \left(\mathcal{M}^{(1)}_{(X,\mathcal{A})} \times \mathcal{F}^{(1)}_{(X,\mathcal{A})}\right) \to [0,1]$ given by*

$$\mathbf{J}_{\mathcal{H}_{(C,n)}}(m,f) = \sup\left\{\sum_{i=1}^{n}\Big(C\big(c_1+\cdots+c_i, m(f \geq c_1+\cdots+c_i)\big)\right.$$
$$\left. -C\big(c_1+\cdots+c_{i-1}, m(f \geq c_1+\cdots+c_i)\big)\Big)\,\middle|\; c_1,\ldots,c_n \in [0,1], \sum_{i=1}^{n} c_i \leq 1\right\}$$

(with convention $c_1 + c_0 = 0$) is a universal integral on $[0,1]$.

It is not difficult to check that each integral $\mathbf{J}_{(C,n)}$ is related by (U2') to the copula C, and that

$$\mathbf{J}_{(C,1)} = \mathbf{J}_C \leq \mathbf{J}_{(C,2)} \leq \cdots \leq \mathbf{J}_{(C,n)} \leq \cdots \leq \mathbf{J}_{(C,\infty)},$$

where $\mathbf{J}_{(C,\infty)} = \sup\{\mathbf{J}_{(C,n)}\}$. In general, $\mathbf{J}_{(C,\infty)} \leq \mathbf{J}_{(C)}$, see (17).

For the basic copulas M and Π it holds:

$$\mathbf{J}_{(M,n)} = \mathbf{Su}$$

is the Sugeno integral, independently of $n \in \mathbb{N}$, and thus also $\mathbf{J}_{(M,\infty)} = \mathbf{Su}$;

$$\mathbf{J}_{(\Pi,1)} = \mathbf{Sh} < \mathbf{J}_{(\Pi,2)} < \cdots < \mathbf{J}_{(\Pi,n)} < \cdots < \mathbf{J}_{(\Pi,\infty)} = \mathbf{Ch},$$

i.e., the limit member $\mathbf{J}_{(\Pi,\infty)}$ is just the Choquet integral.

However, for the third basic copula $W : [0,1]^2 \to [0,1]$ given by $W(u,v) = \max\{u+v-1, 0\}$,

$$\mathbf{J}_{(W,1)} < \mathbf{J}_{(W,2)} < \cdots < \mathbf{J}_{(W,n)} < \cdots < \mathbf{J}_{(W,\infty)} < \mathbf{J}_{(W)}.$$

To see this fact, it is enough to consider the framework of Example 1, where $\mathbf{J}_{(W,1)}(\lambda, f) = \cdots = \mathbf{J}_{(W,\infty)}(\lambda, f) = 0$ but $\mathbf{J}_{(W)} = 1$.

5 Concluding Remarks

We have discussed several integrals considering monotone measures. We have focused only on two major classes of integrals, namely on decomposition integrals (and their generalizations) and on universal integrals. In all discussed approaches, the role of integral sums was stressed.

We have omited several particular approaches to integration in the framework $\bigcup_{(X,\mathcal{A})\in\mathcal{S}} \left(\mathcal{M}_{(X,\mathcal{A})} \times \mathcal{F}_{(X,\mathcal{A})}\right)$. For an extensive overview we recommend the handbook [22], the monographs [33,34] and the overview papers [24,25,26].

For particular integrals, we can recall their axiomatic characterization, such as charaterization of the Lebesgue integral through linear functionals [4]. The Choquet integral is characterized by the comonotone additivity [27], or, equivalently, by the positive homogeneity and horizontal additivity,

$$\mathbf{Ch}(m,f) = \mathbf{Ch}(m, \min\{c, f\}) + \mathbf{Ch}(m, \max\{0, f-c\})$$

for any fixed constant $c \in]\,0, \infty\,[$, see [29,2].
Similarly, the Sugeno integral can be characterized by means of the min - homogeneity and comonotone maxitivity [13]. For more details on axiomatic approach to integration we recommend [7].

Several approaches to decision procedures are based on combination of integrals. These methods are either multistep integrals of the same type, such as multistep Choquet integral [17,19], or they combine different types of integrals, such as the Choquet and the Sugeno integrals [20].

Acknowledgment. The work on this chapter was supported by grant APVV-0073-10 and VEGA1/0171/12.

References

1. Bassan, B., Spizzichino, F.: Relations among univariate aging, bivariate aging and dependence for exchangeable lifetimes. J. Multivariate Anal. 93, 313–333 (2005)
2. Benvenuti, P., Mesiar, R., Vivona, D.: Monotone set functions-based integrals. In: Pap, E. (ed.) Handbook of Measure Theory, vol. II, pp. 1329–1379 (2002)
3. Choquet, G.: Theory of capacities. Ann. Inst. Fourier 5, 131–295 (1953/1954)
4. Dunford, N., Schwarz, J.T.: Linear operators. Part 1. General Theory. Interscience Publ., New York (1966)
5. Durante, F., Sempi, C.: Semicopulæ. Kybernetika 43(2), 209–220 (2007)
6. Even, Y., Lehrer, E.: Decomposition-Integral: Unifying Choquet and the Concave Integrals, `http://www.math.tau.ac.il/~lehrer/Papers/decomposition.pdf` (submitted)
7. Klement, E.P., Mesiar, R.: Discrete Integrals and Axiomatically Defined Functionals. Axioms 1(1), 9–20 (2012)
8. Klement, E.P., Mesiar, R., Pap, E.: Measure-based aggregation operators. Fuzzy Sets and Systems 142(1), 3–14 (2004)
9. Klement, E.P., Mesiar, R., Pap, E.: A universal integral as common frame for Choquet and Sugeno integral. IEEE Transactions on Fuzzy Systems 18, 178–187 (2010)
10. Lebesgue, H.: Intègrale, longueur, aire. Habilitation thesis, Université de Paris, Paris (1902)
11. Lehrer, E.: A new integral for capacities. Economic Theory 39, 157–176 (2009)
12. Lehrer, E., Teper, R.: The concave integral over large spaces. Fuzzy Sets and Systems 15, 2130–2144 (2008)
13. Marichal, J.: An axiomatic approach of the discrete Sugeno integral as a tool to aggregate interacting criteria in a qualitative framewor. IEEE Transactions on Fuzzy Systems 9(1), 164–172 (2001)
14. Mesiar, R., Li, J., Pap, E.: Discrete pseudo-integrals. International Journal of Approximate Reasoning 54(3), 357–364 (2013)

15. Mesiar, R., Rybárik, J.: Pan-operations structure. Fuzzy Sets and Systems 74(3), 365–369 (1995)
16. Mesiar, R.: Stupňanová, A.: Decomposition integrals. Int. Journal of Approximate Reasoning (in print, 2013)
17. Mesiar, R., Vivona, D.: Two-step integral with respect to fuzzy measure. Tatra Mt. Math. Publ. 16(pt. II), 359–368 (1999)
18. Mostert, P.S., Shield, A.L.: On the structure of semigroups on a compact manifold with boundary. Ann. of Math. 65, 117–143 (1957)
19. Murofushi, T., Sugeno, M., Fujimoto, K.: Separated hierarchical decomposition of the choquet integral. International Journal of Uncertainty, Fuzziness and Knowledge-Based Systems 5(5), 563–585 (1997)
20. Narukawa, Y., Torra, V.: Twofold integral and Multi-step Choquet integral. Kybernetika 40(1), 39–50 (2004)
21. Nelsen, R.B.: An Introduction to Copulas, 2nd edn. Lecture Notes in Statistics, vol. 139. Springer, New York (2006)
22. Pap, E.: Handbook of Measure Theory, Part I, Part II. Elsevier, Amsterdam (2002)
23. Riemann, B.: Über die Darstellbarkeit einer Function durch eine trigonometrische Reihe (On the representability of a function by a trigonometric series). Habilitation thesis, University of Göttingen (1854)
24. Sander, W., Siedekum, J.: Multiplication, distributivity and fuzzy-integral. I Kybernetika 41(3), 397–422 (2005)
25. Sander, W., Siedekum, J.: Multiplication, distributivity and fuzzy-integral. II Kybernetika 41(4), 469–496 (2005)
26. Sander, W., Siedekum, J.: Multiplication, distributivity and fuzzy-integral. III Kybernetika 41(3), 497–518 (2005)
27. Schmeidler, D.: Integral representation without additivity. Proc. Amer. Math. 97(2), 255–270 (1986)
28. Shilkret, N.: Maxitive measure and integration. Indag. Math. 33, 109–116 (1971)
29. Šipoš, J.: Integral with respect to a pre-measure. Math. Slovaca 29(2), 141–153 (1979)
30. Sugeno, M.: Theory of Fuzzy Integrals and its Applications. PhD thesis, Tokyo Institute of Technology (1974)
31. Sugeno, M., Murofushi, T.: Pseudo-additive measures and integrals. J. Math. Anal. Appl. 122, 197–222 (1987)
32. Stupňanová, A.: A Note on Decomposition Integrals. In: Greco, S., Bouchon-Meunier, B., Coletti, G., Fedrizzi, M., Matarazzo, B., Yager, R.R. (eds.) IPMU 2012, Part IV. CCIS, vol. 300, pp. 542–548. Springer, Heidelberg (2012)
33. Torra, V., Narukawa, Y.: Modeling Decisions: Information Fusion and Aggregation Operators. In: Cognitive Technologies. Springer (2007)
34. Wang, Z., Klir, G.J.: Generalized Measure Theory. Springer, Heidelberg (2009)
35. Yang, Q.: The Pan-integral on the Fuzzy Measure Space. Fuzzy Mathematics 3, 107–114 (1985)
36. Zadeh, L.A.: Fuzzy sets. Information and Control 8, 338–353 (1965)
37. Zadeh, L.A.: Probability measures of fuzzy events. J. Math. Anal. Applic. 23, 421–427 (1968)

Entropy of Capacity

Aoi Honda

Kyushu Institute of Technology, 680-4 Kawazu, Iizuka, Fukuoka 820-8502, Japan

Abstract. Capacity is a monotone set function and an important class of non-additive measure. In this chapter entropies of capacities are discussed. First, entropies for classical capacities are introduced, then they are generalized for capacities on set systems. Moreover axiomatizations of the entropies are given to characterize them.

1 Introduction

The classical definition of the Shannon entropy [13] for a probability measure is at the core of information theory. Therefore, many attempts for defining an entropy for a set function more general than the classical probability measure have been done, in particular for a capacity, which is a special class of non-additive measures and is called a fuzzy measure [15] or a monotone set function.

This chapter is organized as follows. In Section 2, we introduce three definitions of entropy for a capacity, Yager's [16], Marichal and Roubens's [12], and Dukhovny's [3] and compare their properties. In Section 3, we generalize these three entropies for the sake of making them applicable to more general capacities, which are defined on subsets of the power set, not on the whole power set. In Section 4 we show that multi-choice game and bi-capacity can be regarded as capacities on set systems. Then they can be treated in our framework and that enables us to apply the entropies to them. In Section 5 we show the axiomatizations of generalized Marichal and Roubens's entropy and Dukhovny's minimum entropy.

Throughout this chapter, we assume that the whole set is $N = \{1, 2, \dots, n\}$.

2 Entropies of Classical Capacity

We start this section devoted to several definitions of entropy for capacities reviewing a few basic definitions needed later on.

Definition 1 (classical capacity). *A set function* $v : 2^N \to [0, 1]$ *is a* capacity *on* N *if it satisfies* $v(\emptyset) = 0, v(N) = 1$ *and* $v(A) \leq v(B)$ *whenever* $A \subseteq B$ *for any* $A, B \in 2^N$.

Probability measure on $(N, 2^N)$ is a set function which satisfies $p(\emptyset) = 0$, $p(N) = 1$ and for any $A, B \in 2^N$ $p(A \cup B) = p(A) + p(B)$ whenever $A \cap B = \emptyset$.

The Shannon entropy of a probability measure p is defined as follows.

V. Torra, Y. Narukawa, and M. Sugeno (eds.), *Non-Additive Measures*,
Studies in Fuzziness and Soft Computing 310,
DOI: 10.1007/978-3-319-03155-2_4,

Definition 2 (Shannon entropy [12]). *Let p be a probability measure on $(N, 2^N)$. Then the Shannon entropy of p is defined by*

$$H_{\mathrm{S}}(p) := \sum_{i=1}^{n} h(p(\{i\})), \tag{1}$$

where

$$h(x) := -x \log x. \tag{2}$$

Yager [16] and Marichal-Roubens [12] proposed definitions for a capacity having desirable properties, and which can be considered as generalizations of the Shannon entropy.

Definition 3 (Yager's entropy [16]). *Let v be a capacity on $(N, 2^N)$. Then the Yager's entropy of v is defined by*

$$H_{\mathrm{Y}}(v) := \sum_{i=1}^{n} h\left(\sum_{A \subseteq N \setminus \{i\}} \gamma^n_{|A|} (v(A \cup \{i\}) - v(A)) \right), \tag{3}$$

where

$$\gamma^n_k := \frac{(n-k-1)!k!}{n!}. \tag{4}$$

Definition 4 (Marichal-Roubens's entropy [12]). *Let v be a capacity on $(N, 2^N)$. Then the Marichal-Roubens's entropy of v is defined by*

$$H_{\mathrm{MR}}(v) := \sum_{i=1}^{n} \sum_{A \subseteq N \setminus \{i\}} \gamma^n_{|A|} h(v(A \cup \{i\}) - v(A)). \tag{5}$$

A difference between Yager's and Marichal-Roubens's entropy is only the place of the function h.

Another attempt was also done by Dukhovny [3], which is called a minimum entropy. To describe the definition we introduce a concept of the maximum chain needed in a definition of the minimum entropy. For $a, b \in 2^N$, we say a is covered by b, and write $a \prec b$ or $b \succ a$, if $a \subset b$ and $a \subset x \subsetneq b$ implies $x = a$. In other words, $a \prec b$ means that there are no elements between a and b.

Definition 5 (maximal chain of 2^N). $\mathcal{C} = (c_0, c_2, \ldots, c_n), c_i \in 2^N, i = 1, \ldots n$ *is called a maximal chain of 2^N if $\mathcal{C}$ satisfies that $\emptyset = c_0 \prec c_1 \prec \cdots \prec c_n = N$.*

We denote the set of all maximal chains of 2^N by $\mathcal{M}(2^N)$. Let $\mathcal{C}$ be a maximal chain of 2^N. We define $p^{v,\mathcal{C}} = (p_1^{v,\mathcal{C}}, p_2^{v,\mathcal{C}}, \ldots, p_n^{v,\mathcal{C}})$ by

$$\begin{aligned} p^{v,\mathcal{C}} &= (p_1^{v,\mathcal{C}}, p_2^{v,\mathcal{C}}, \ldots, p_n^{v,\mathcal{C}}) \\ &:= (v(c_1) - v(c_0), v(c_2) - v(c_1), \ldots, v(c_n) - v(c_{n-1})), \end{aligned} \tag{6}$$

Note that $p^{v,\mathcal{C}}$ satisfies $p_i^{v,\mathcal{C}} \geq 0, i = 1, \ldots, n$ and $\sum_{i=1}^{n} p_i^{v,\mathcal{C}} = 1$ like a probability measure.

Definition 6 (minimum entropy [3]). *Let v be a capacity on $(N, 2^N)$. The minimum entropy of v is defined by*

$$H_{\min}(v) := \min_{\mathcal{C}\in\mathcal{M}(2^N)} \sum_{i=1}^{n} h(p_i^{v,\mathcal{C}}) = \min_{\mathcal{C}\in\mathcal{M}(2^N)} H_{\mathrm{S}}(p^{v,\mathcal{C}}). \tag{7}$$

Example 1. Define v_A, v_B and v_c on $(N = \{1,2,3\}, 2^N)$ as Table 1. Then each entropy of v_A, v_B and v_C is as Table 2. Remark that $1.5850 = \log_2 3$.

Table 1. capacities

	$\emptyset$	$\{1\}$	$\{2\}$	$\{3\}$	$\{1,2\}$	$\{1,3\}$	$\{2,3\}$	N
v_A	0.0	1/3	1/3	1/3	2/3	2/3	2/3	1.0
v_B	0.0	0.8	0.8	0.8	0.9	0.9	0.9	1.0
v_C	0.0	0.2	0.3	0.4	0.8	0.7	0.6	1.0

Table 2. Entropies

	$H_{\mathrm{Y}}(\,\cdot\,; v_T)$	$H_{\mathrm{MR}}(\,\cdot\,; v_T)$	$H_{\min}(\,\cdot\,; v_T)$
v_A	1.5850	1.5850	1.5850
v_B	1.5850	0.9219	0.9219
v_C	1.5850	1.5010	1.3710

Marichal-Roubens's entropy can be represented using maximal chain.

Proposition 1 (representation of Marichal-Roubens's entropy). *Let v be a capacity on $(N, 2^N)$. Then we have*

$$H_{\mathrm{MR}}(v) = \frac{1}{|n!|} \sum_{\mathcal{C}\in\mathcal{M}(2^N)} H_{\mathrm{S}}(p^{v,\mathcal{C}}). \tag{8}$$

Remark that $|\mathcal{M}(2^N)| = n!$ Proposition 14 shows that H_{MR} is obtained by replacing the minimum operation of $H_{\min}$ with the average operator.

On the other hand, Yager's entropy can be represented using the Shapley value which is the most important concept in game theory.

Definition 7 (Shapley value). *Let v be a capacity on $(N, 2^N)$. The Shapley value of v, $\Phi(v) := (\phi_1(v), \ldots, \phi_n(v)) \in [0,1]^n$ is defined by*

$$\phi_i(v) := \sum_{A\subseteq N\setminus\{i\}} \gamma_{|A|}^{n} [v(A\cup\{i\}) - v(A)]. \tag{9}$$

Remark that $\sum_{i=1}^{n} \phi_i(v) = 1$ holds.

Proposition 2 (representation of Yager's entropy). *Let v be a capacity on $(N, 2^N)$. Then we have*

$$\begin{aligned} H_{\mathrm{Y}}(v) &= \sum_{i=1}^{n} h(\phi_i(v)) \\ &= H_{\mathrm{S}}(\Phi(v)). \end{aligned} \tag{10}$$

Next we study properties of these entropies.

Proposition 3. *Let v be a capacity on $(N, 2^N)$.*
(i) *If v is an additive function, then $H_{\mathrm{Y}}(v) = H_{\mathrm{MR}}(v) = H_{\min}(v) = H_{\mathrm{S}}(v)$.*
(ii) *All $H_{\mathrm{Y}}(v), H_{\mathrm{MR}}(v)$ and $H_{\min}(v)$ are continuous functions.*
(iii) *$H_{\min}(v)$ is not a differentiable function.*
(iv) *All $H_{\mathrm{Y}}(v), H_{\mathrm{MR}}(v)$ and $H_{\min}(v)$ lead to a value between $[0, \log n]$.*
(v) *If v is a $\{0,1\}$-valued capacity, then all $H_{\mathrm{Y}}(v), H_{\mathrm{MR}}(v)$ and $H_{\min}(v)$ lead to a value 0. The converse does not holds except for $H_{\mathrm{MR}}(v)$.*
(vi) *All H_{Y}, H_{MR}, $H_{\min}$ take $\log_2 n$ as the max value.*
(vii) *If $v = v^*(A) := |A|/n, A \in 2^N$. Then $H_{\mathrm{Y}}(v), H_{\mathrm{MR}}(v)$ and $H_{\min}(v)$ is $\log_2 n$. The converse holds except for $H_{\mathrm{Y}}(v)$.*

Proposition 4. *Define $v_\lambda := (1-\lambda)v + \lambda v^*, 0 < \lambda < 1$. For any $v(\not\equiv v^*)$, $0 < \lambda_1 < \lambda_2 < 1$ implies $H_{\mathrm{Y}}(v_{\lambda_1}) \leq H_{\mathrm{Y}}(v_{\lambda_2})$, $H_{\mathrm{MR}}(v_{\lambda_1}) < H_{\mathrm{MR}}(v_{\lambda_2})$ and $H_{\min}(v_{\lambda_1}) < H_{\min}(v_{\lambda_2})$.*

In other words, $H_{\mathrm{MR}}(v_\lambda)$ and $H_{\min}(v_\lambda), 0 < \lambda < 1$ are strictly increasing functions of λ. Concerning to H_{Y}, it is an increasing function of λ, but not strictly increasing (Cf. Prop 3 (vi)).

We show the graphs of these three entropies in the case of $n = 2$, that is, N is a two point set $N = \{1, 2\}$. On the left side, we display the 3D graphs and on the right side, we display their contour graphs (Fig 1-3). All sections of these graphs have been created by a plane cutting to $v(\{1\}) + v(\{2\}) = 1$ which coincide with the graph of Shannon entropy (Cf. Prop 3 (i)).

We can also generalized the relative entropy, called also the Kullback-Leibler divergence, in the above three ways.

Definition 8 (relative entropy). *Let p and q be probability measures on N. The relative entropy of p to q is defined by*

$$H_{\mathrm{KL}}(p; q) := \sum_{i=1}^{n} h[p(\{i\}), q(\{i\})], \tag{11}$$

where $h(x; y) := x \log(x/y)$.

Definition 9 (relative entropy H_{Y}). *Let v and u be capacities on $(N, 2^N)$. $H_{\mathrm{Y}}(v; u)$ is defined by*

$$H_{\mathrm{Y}}(v; u) := H_{\mathrm{KL}}(\Phi(v); \Psi(u)). \tag{12}$$

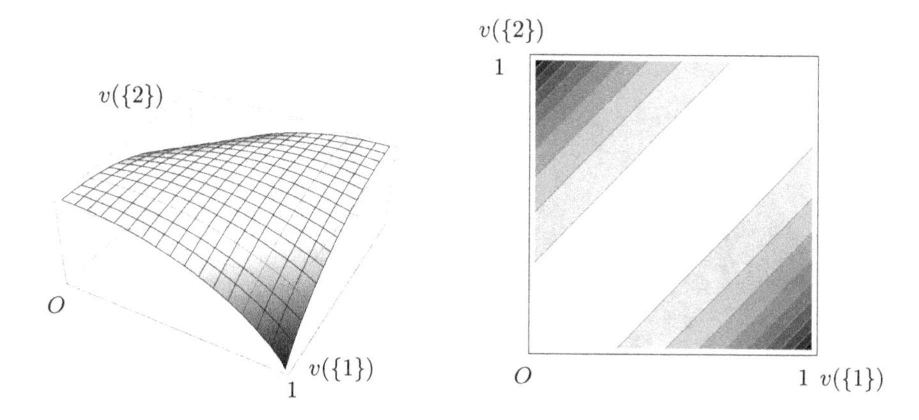

Fig. 1. The graph of $H_{\mathrm{Y}}(v)$ in the case of $N = \{1, 2\}$

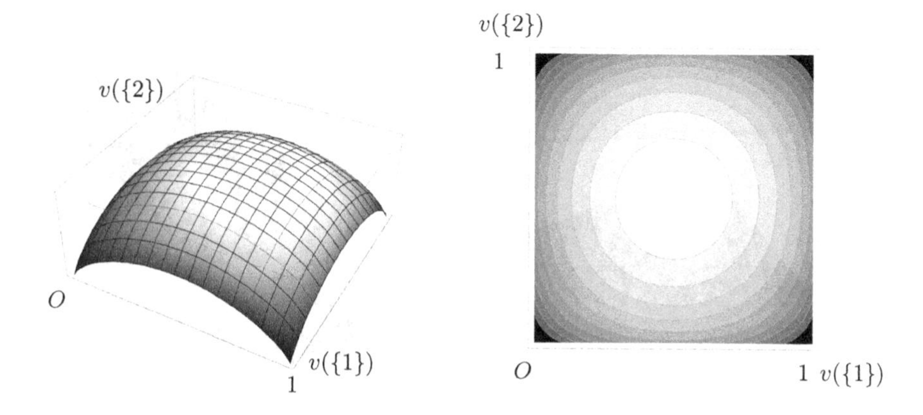

Fig. 2. The graph of $H_{\mathrm{MR}}(v)$ in the case of $N = \{1, 2\}$

Definition 10 (relative entropy H_{MR}). *Let v and u be capacities on $(N, 2^N)$. $H_{\mathrm{MR}}(v; u)$ is defined by*

$$H_{\mathrm{MR}}(v; u) := \frac{1}{|\mathcal{M}(2^N)|} \sum_{\mathcal{C} \in \mathcal{M}(\mathcal{S})} H_{\mathrm{KL}}(p^{v,\mathcal{C}}; p^{u,\mathcal{C}}). \tag{13}$$

Definition 11 (relative entropy $H_{\min}$). *Let v and u be capacities on $(N, 2^N)$. $H_{\min}(v; u)$ is defined by*

$$H_{\min}(v; u) := \min_{\mathcal{C} \in \mathcal{M}(2^N)} H_{\mathrm{KL}}(p^{v,\mathcal{C}}; p^{u,\mathcal{C}}). \tag{14}$$

We show the graphs of these three relative entropies of v defined on $(\{1, 2\}, 2^{\{1,2\}})$ to $u(\{1\}) = 0.2, u(\{2\}) = 0.3$ in the case of $N = \{1, 2\}$. On the left side, we display the 3D graphs and on the right side we display their contour graphs. (Fig 4-6).

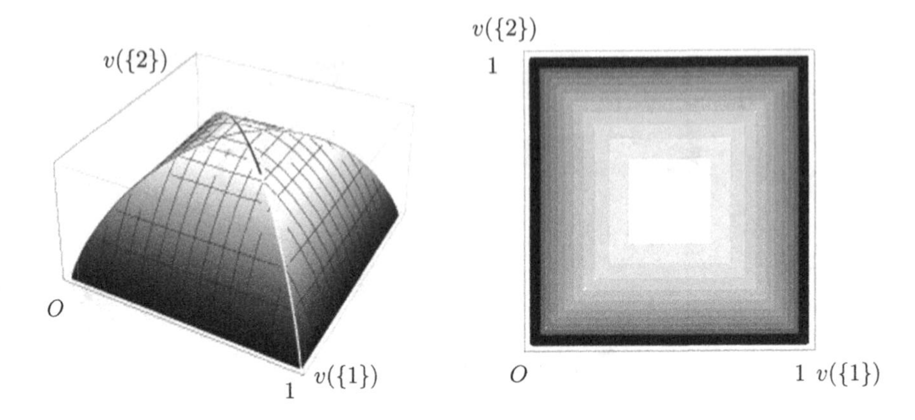

Fig. 3. The graph of $H_{\min}(v)$ in the case of $N = \{1, 2\}$

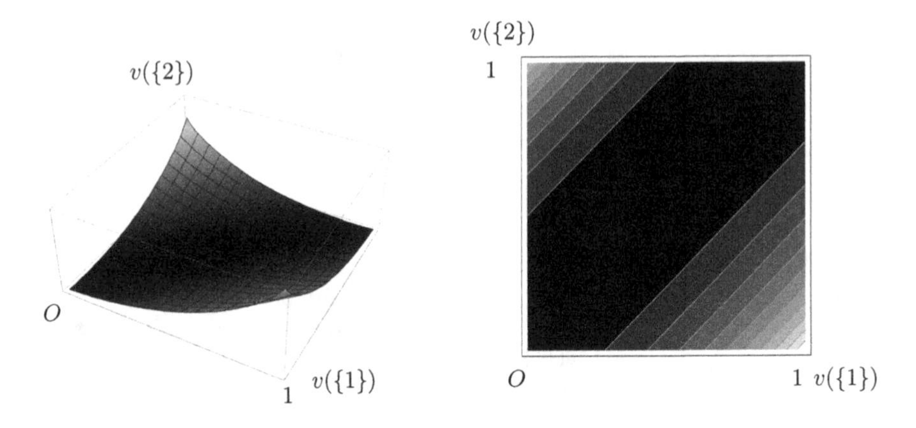

Fig. 4. The graph of $H_{\mathrm{Y}}(v; u)$ in the case of $N = \{1, 2\}$

Example 2. Define v_T, v_A, v_B and v_c on $(N = \{1, 2, 3\}, 2^N)$ as Table 3. Then the mean squared error of v_T and other capacities has the same value, 0.04. However relative entropies of v_A, v_B and v_C to v_T are as Table 4.

3 Entropy of Capacity on Set System

In this section, we consider more general capacities, in the sense that the underlying system of sets may be not 2^N. Let $\mathcal{S}$ be a subset of 2^N. We call $(N, \mathcal{S})$, or simply $\mathcal{S}$, a *set system* if $\emptyset, N \in \mathcal{S}$. Then the capacity is naturally generalized on $\mathcal{S}$.

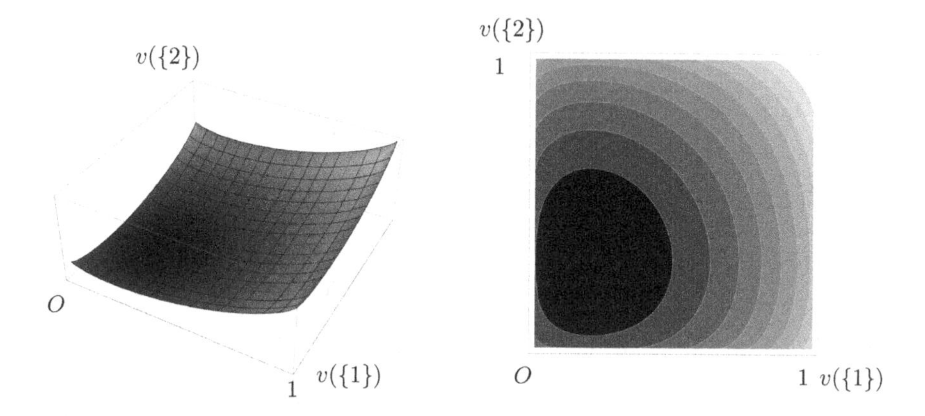

Fig. 5. The graph of $H_{\mathrm{MR}}(v;u)$ in the case of $N = \{1,2\}$

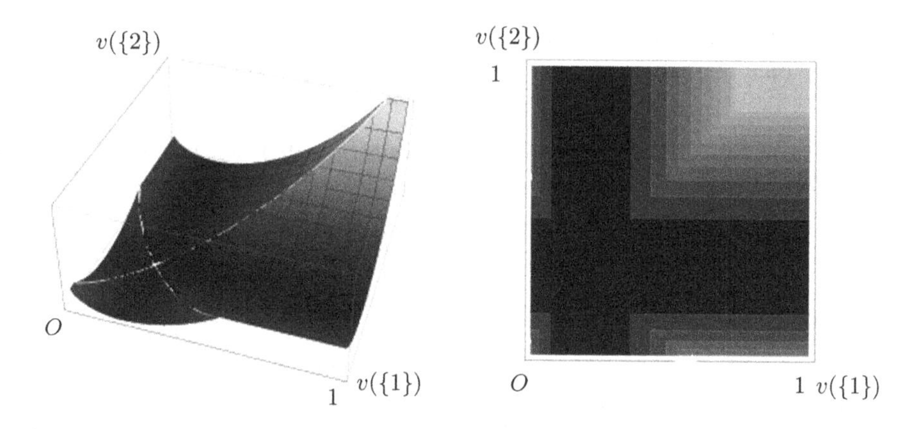

Fig. 6. The graph of $H_{\min}(v;u)$ in the case of $N = \{1,2\}$

Definition 12 (capacity on set system). *Let $(N,\mathcal{S})$ be a set system. A set function $v : \mathcal{S} \to [0,1]$ is a* capacity on $(N,\mathcal{S})$ *if it satisfies $v(\emptyset) = 0, v(N) = 1$ and $v(A) \leq v(B)$ whenever $A \subseteq B$ for any $A, B \in \mathcal{S}$.*

The maximal chain of $\mathcal{S}$ is defined in the same manner as for the case of 2^N. We introduce a kind of the regularity on a set system.

Definition 13 (regular set system). *Let $(N,\mathcal{S})$ be a set system. We say that $(N,\mathcal{S})$ is a regular set system if for any $\mathcal{C} \in \mathcal{M}(\mathcal{S})$, the length of $\mathcal{C}$ is n, i.e. $|\mathcal{C}| = n+1$.*

In other words, for any A that belongs to a regular set system $(N,\mathcal{S})$, there is $i \in N$ satisfying $A \cup \{i\} \in \mathcal{S}$ and $i \notin A$.

Table 3. capacities

	$\emptyset$	$\{1\}$	$\{2\}$	$\{3\}$	$\{1,2\}$	$\{1,3\}$	$\{2,3\}$	N
v_T	0.0	0.3	0.3	0.3	0.7	0.7	0.7	1.0
v_A	0.0	0.3	0.4	0.4	0.8	0.8	0.7	1.0
v_B	0.0	0.3	0.4	0.4	0.7	0.6	0.6	1.0
v_C	0.0	0.2	0.3	0.4	0.8	0.7	0.6	1.0

Table 4. Relative entropies

	$H_Y(\,\cdot\,; v_T)$	$H_{MR}(\,\cdot\,; v_T)$	$H_{\min}(\,\cdot\,; v_T)$
v_A	0.0000	0.0448	0.0415
v_B	0.0036	0.0798	0.0000
v_C	0.0000	0.0700	0.0415

All of the three entropies introduced in Section 2 can be generalized for a capacity on a regular set system. Dukhovny's entropy is generalized in a natural way.

Definition 14 (Generalized minimum entropy [9]). *Let v be a capacity on a regular set system $(N, \mathcal{S})$. Then $H_{\min}(v)$ is defined by*

$$H_{\min}(v) := \min_{\mathcal{C} \in \mathcal{M}(\mathcal{S})} H_S(p^{v,\mathcal{C}}). \tag{15}$$

Marichal-Roubens's entropy is generalized using its representation in Proposition 14.

Definition 15 (Generalized Marichal-Roubens's entropy [8]). *Let v be a capacity on a regular set system $(N, \mathcal{S})$. Then $H_{MR}(v)$ is defined by*

$$H_{MR}(v) := \frac{1}{|\mathcal{M}(\mathcal{S})|} \sum_{\mathcal{C} \in \mathcal{M}(\mathcal{S})} H_S(p^{v,\mathcal{C}}). \tag{16}$$

To generalize Yager's entropy for a capacity on a regular set system, we first generalize the Shapley value.

Definition 16 (Shapley value [14,1,5])
Let v be a capacity on a regular set system $(N, \mathcal{S})$. The Shapley value of v, $\Phi(v) := (\phi_1(v), \ldots, \phi_n(v)) \in [0,1]^n$ is defined by

$$\phi_i(v) := \frac{1}{|\mathcal{M}(\mathcal{S})|} \sum_{\substack{\mathcal{C} \in \mathcal{M}(\mathcal{S}) \\ A, A\{i\} \in \mathcal{C}}} [v(A \cup \{i\}) - v(A)]. \tag{17}$$

Yager's entropy is generalized using its representation in Proposition 2 and the generalized Shapley value.

Definition 17 (Generalized Yager's entropy). *Let v be a capacity on a regular set system $(N, \mathcal{S})$. $H_{\mathrm{MR}}(v)$ is defined by*

$$H_{\mathrm{Y}}(v) := H_{\mathrm{S}}(\Phi(v)). \tag{18}$$

We can also generalize the relative entropy for a capacity on a regular set system.

Definition 18 (Generalized relative entropy H_{Y}). *Let v and u be capacities on a regular set system $(N, \mathcal{S})$. $H_{\mathrm{Y}}(v; u)$ is defined by*

$$H_{\mathrm{Y}}(v; u) := H_{\mathrm{KL}}(\Phi(v); \Psi(u)). \tag{19}$$

Definition 19 (Generalized relative entropy H_{MR}). *Let v and u be capacities on a regular set system $(N, \mathcal{S})$. $H_{\mathrm{MR}}(v; u)$ is defined by*

$$H_{\mathrm{MR}}(v; u) := \frac{1}{|\mathcal{M}(\mathcal{S})|} \sum_{\mathcal{C} \in \mathcal{M}(\mathcal{S})} H_{\mathrm{KL}}(p^{v,\mathcal{C}}; p^{u,\mathcal{C}}). \tag{20}$$

Definition 20 (Generalized relative entropy $H_{\min}$). *Let v and u be capacities on a regular set system $(N, \mathcal{S})$. $H_{\min}(v; u)$ is defined by*

$$H_{\min}(v; u) := \min_{\mathcal{C} \in \mathcal{M}(\mathcal{S})} H_{\mathrm{KL}}(p^{v,\mathcal{C}}; p^{u,\mathcal{C}}). \tag{21}$$

4 Representation as Set System

The genralized entropies presented in Section 3 have applicability to capacities defined on regular set systems. In this section, we show that almost all capacities which appear in applications can be regarded as capacities on regular set systems(cf. [8]).

Let $(L, \leq)$ be a lattice, that is, $(L, \leq)$ is a partially ordered set such that for any pair $x, y \in L$ there exist the least upper bound $x \vee y$ (supremum) and the greatest lower bound $x \wedge y$ (infimum) in L. Consequently, for finite lattices, there always exist the greatest element (supremum of all elements) and the least element (infimum of all elements), denoted by $\top, \bot$ (see [2]).

Definition 21 (capacity on lattice). *Let $(L, \leq)$ be a finite lattice with the greatest and the least elements denoted by $\top$ and $\bot$ respectively. A* capacity *on L is a function $v : L \to [0, 1]$ satisfying $v(\bot) = 0$, $v(\top) = 1$, and being monotone, i.e., $x \leq y$ implies $v(x) \leq v(y)$.*

Evidently a set system is not necessarily a lattice. Moreover, a regular set system is not necessarily a lattice. Indeed, take $N = \{1, 2, 3, 4\}$ and $\mathcal{S} := \{\emptyset, \{1\}, \{3\}, \{1, 2\}, \{2, 3\}, \{14\}, \{3, 4\}, \{1, 2, 3\}, \{1, 3, 4\}, N\}$. Then, $\{1\}$ and $\{3\}$ have no supremum.

Definition 22 (join-irreducible element). *An element $x \in (L, \leq)$ is* join-irreducible *if for all $a, b \in L$, $x \neq \bot$ and $x = a \vee b$ implies $x = a$ or $x = b$.*

We denote $\mathcal{J}(L)$ the set of all join-irreducible elements of L.

The mapping η for any $a \in L$, defined by

$$\eta(a) := \{x \in \mathcal{J}(L) \mid x \leq a\}$$

is a lattice-isomorphism of L onto $\eta(L) := \{\eta(a) \mid a \in L\}$, that is, $(L, \leq) \cong (\eta(L), \subseteq)$ (Figure 7). Clearly $(\mathcal{J}(L), \eta(L))$ is a set system.

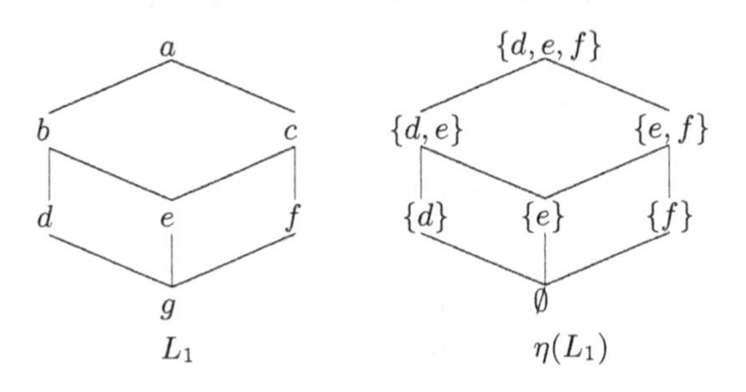

Fig. 7. Translation of lattice

Translating a lattice, which is an underlying space of a capacity v, to a set systems we can apply the entropies introduced in Section 3 to a capacity if the underlying set system is regular.

Remark 1. If we regard $\mathcal{S} := \{\emptyset, \{1\}, \{1,2,3\}\}$ as a lattice, not as a set system, the situation is a little different. In this case, the name of the elements are just the label. We have $\mathcal{J}(\mathcal{S}) = \{\{1\}, \{1,2,3\}\}$ and $|\mathcal{J}(\mathcal{S})| = 2$, so that considering $\mathcal{S}$ as a lattice, we can translate it to a regular set system.

Lemma 1. *If* $(L, \leq)$ *satisfies the following property:*

($\vee$-minimal regular) *for any* $C \in \mathcal{C}(L)$, *the length of* C *is* $|\mathcal{J}(L)|$, *i.e.* $|C| = |\mathcal{J}(L)| + 1$,

then $(\mathcal{J}(L), \eta(L))$ *is a regular set system.*

The maximal chain $C = (c_0, c_1, \ldots, c_m)$ of a lattice is also defined by $\perp = c_0 \prec c_1 \prec \cdots \prec c_m = \top, c_i \in \mathcal{S}, i = 0, \ldots, m.$ in the same way of a set system, where $A \prec B$ denotes that $A \leq B$ and $A \leq C < B$ implies $C = A$.

Most lattices which underlie capacities appearing in practice, such as bi-capacities, and multi-choice games are $\vee$-minimal regular; therefore the entropies are applicable to these cases. We show several practical examples of capacities.

- *Regular lattice and its translation to a set system*

L_1 in Figure 7 is a $\vee$-minimal regular lattice, and $\eta(L_1)$ in Figure 7 is the translation of a lattice L_1 to a set system.

In fact, $\mathcal{J}(L_1) = \{d, e, f\}$, and L_1 is also represented by $\eta(L_1)$. $\mathcal{C}(\eta(L_1)) = \{(\emptyset, \{\mathrm{d}\}, \{\mathrm{d,e}\}, \{\mathrm{d,e,f}\}), (\emptyset, \{\mathrm{e}\}, \{\mathrm{d,e}\}, \{\mathrm{d,e,f}\}), (\emptyset, \{\mathrm{e}\}, \{\mathrm{e,f}\}, \{\mathrm{d,e,f}\}), (\emptyset, \{\mathrm{f}\}, \{\mathrm{e,f}\}, \{\mathrm{d,e,f}\})\}$. Let v be a capacity on L_1. Then the entropy of v on L_1 is

$$\begin{aligned}H(v) = &\frac{1}{4}h[v(d) - v(g)] + \frac{1}{2}h[v(e) - v(g)] + \frac{1}{4}h[v(f) - v(g)]\\ &+\frac{1}{4}h[v(b) - v(d)] + \frac{1}{4}h[v(b) - v(e)] + \frac{1}{4}h[v(c) - v(e)]\\ &+\frac{1}{4}h[v(c) - v(f)] + \frac{1}{2}h[v(a) - v(b)] + \frac{1}{2}h[v(a) - v(c)].\end{aligned}$$

• *Bi-capacity* [6][7]
A bi-capacity is a monotone function on $\mathcal{Q}(N) := \{(A, B) \in 2^N \times 2^N \mid A \cap B = \emptyset\}$ which satisfies $v(\emptyset, N) = -1$, $v(\emptyset, \emptyset) = 0$ and $v(N, \emptyset) = 1$. For any $(A_1, A_2), (B_1, B_2) \in \mathcal{Q}(N)$, $(A_1, A_2) \sqsubseteq (B_1, B_2)$ iff $A_1 \subseteq B_1$ and $A_2 \supseteq B_2$. $\mathcal{Q}(N) \cong 3^N$. It can be shown that $(\mathcal{Q}(N), \sqsubseteq)$ is a finite distributive lattice. Sup and inf are given by $(A_1, A_2) \vee (B_1, B_2) = (A_1 \cup B_1, A_2 \cap B_2)$ and $(A_1, A_2) \wedge (B_1, B_2) = (A_1 \cap B_1, A_2 \cup B_2)$, and we have

$$\mathcal{J}(\mathcal{Q}(N)) = \{(\emptyset, N \setminus \{i\}), i \in N\} \cup \{(\{i\}, N \setminus \{i\}), i \in N\},$$

where $i \in N$. Normalizing v by $v' : \mathcal{Q}(N) \to [0, 1]$ such that

$$v' := \frac{1}{2}v + \frac{1}{2},$$

$$\begin{aligned}H(v') = \sum_{i=1}^{n} \sum_{\substack{A \subset N \setminus x_i \\ B \subset N \setminus (A \cup \{i\})}} \gamma^n_{|A|,|B|} &\left(h\left[v'(A \cup \{i\}, B) - v'(A, B)\right]\right.\\ &\left.+h\left[v'(B, A) - v'(B, A \cup \{i\})\right]\right).\end{aligned}$$

where

$$\gamma^n_{k,\ell} := \frac{(n - k + \ell - 1)!\,(n + k - \ell)!\,2^{n-k-\ell}}{(2n)!}.$$

• *Multi-choice game*
Let $N := \{0, 1, \ldots .n\}$ be a set of players, and let $L := L_1 \times \cdots \times L_n$, where $(L_i, \leq_i)$ is a totally ordered set $L_i = \{0, 1, \ldots, \ell_i\}$ such that $0 \leq_i 1 \leq_i \cdots \leq_i \ell_i$. Each L_i is the set of choices of player i. $(L, \leq)$ is a regular lattice. For any $(a_1, a_2, \ldots, a_n), (b_1, b_2, \ldots, b_n) \in L$, $(a_1, a_2, \ldots, a_n) \leq (b_1, b_2, \ldots, b_n)$ iff $a_i \leq_i b_i$ for all $i = 1, \ldots, n$. We have

$$\mathcal{J}(L) = \{(0, \ldots, 0, a_i, 0, \ldots, 0) \mid a_i \in \mathcal{J}(L_i) = L_i \setminus \{0\}\}$$

and $|\mathcal{J}(L)| = \sum_{i=1}^{n} \ell_i$. The lattice in Fig. 8 is an example of a product lattice, which represents a 2-player game. Players 1 and 2 can choose among 3

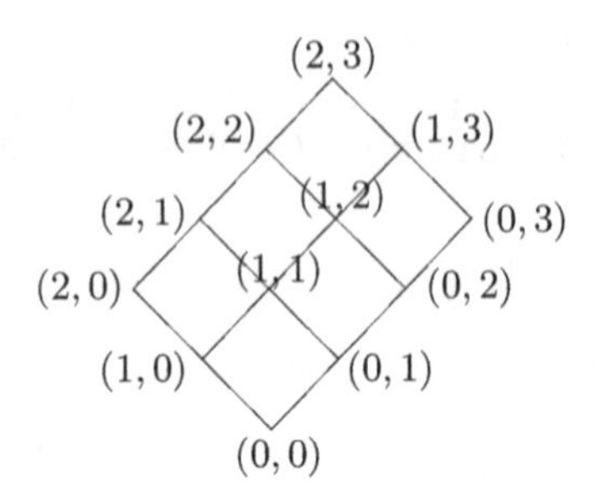

Fig. 8. 2-players game

and 4 choices, respectively. Let v be a capacity on L, that is, $v(0,\dots,0)=0$, $v(\ell_1,\dots,\ell_n)=1$ and , for any $a,b\in L$, $v(a)\leqq v(b)$ whenever $a\le b$. In this case, we have

$$H(v)=\sum_{\substack{i\in N\\ j\in L_i}}\sum_{a\in L/L_i}\xi_i^{(a,j)}\ h\left[v(a,j)-v(a,j-1)\right]$$

where $L/L_i:=L_1\times\cdots\times L_{i-1}\times L_{i+1}\times\cdots\times L_n$, $(a,a_i):=(a_1,\dots,a_{i-1},a_i,a_{i+1},\dots,a_n)\in L$ such that $a\in L/L_i$ and $a_i\in L_i$,

$$\xi_i^{(a,a_i)}:=\left(\prod_{k=1}^{n}\binom{\ell_k}{a_k}\right)\cdot\binom{\sum_{k=1}^n \ell_k}{\sum_{k=1}^n a_k}^{-1}\cdot\frac{a_i}{\sum_{k=1}^n a_k}.$$

5 Axiomatization of Entropy of Capacity

In this section, we discuss the characterizations of entropies of capacities focusing with axiomatizations.

Before introducing the axioms for an entropy of a capacity, we discuss about the domains of an entropy. Let $p:=(p_1,\dots,p_n)$ be a probability measure on $N:=\{1,2,\dots,n\}$. Then $(N,2^N,p)$ is called a probability space. Let Δ_n be the set of all probability measure on N. H_{S} is a function defined on $\Delta:=\bigcup_{n=1}^{\infty}\Delta_n$ to $[0,\infty)$. We should denote $H_{\mathrm{S}}(p)$ with the underlying space by $H_{\mathrm{S}}(N,2^N,p)$, and as far as no confusion occurs we denote $H_{\mathrm{S}}(N,2^N,p)$ simply $H_{\mathrm{S}}(p)$.

Similarly, let v be a capacity on $(N,\mathcal{S})$. Then we call $(N,\mathcal{S},v)$ a capacity space. Let Σ_n be the set of all regular set systems of $N:=\{1,2,\dots,n\}$ and let $\Delta'_{\mathcal{S}}$ be the set of all capacity spaces defined on the regular set system $(N,\mathcal{S})$. The domain of the entropy of the capacity is $\Delta':=\bigcup_{n=1}^{\infty}\bigcup_{\mathcal{S}\in\Sigma_n}\Delta'_{\mathcal{S}}$ and the entropy is a function defined on Δ' to $[0,\infty)$. We denote simply $H(v)$ instead of $H(N,\mathcal{S},v)$ as far as no ambiguity occurs. More properly, the dual capacity of v is the dual capacity space of the capacity space $(N,\mathcal{S},v)$ which is defined by $(N,\mathcal{S},v)^d:=(N,\mathcal{S}^d,v^d)$ with $\mathcal{S}^d:=\{A^c\in 2^N\mid A\in\mathcal{S}\}$, the permutation of v is the permutation of the capacity space $(N,\mathcal{S},v)$ which is defined by $(N,\mathcal{S},v)^\pi:=(N,\pi(\mathcal{S}),\pi\circ v)$, and the embedding of $v^{\mathbf{2}}$ is the embedding of the capacity space $(N,\mathbf{2},v^{\mathbf{2}})$ which is defined by $(N,\mathcal{S},v)^{c_k}:=(N^{c_k},\mathcal{S}^{c_k},v^{c_k})$.

First, we show Faddeev's axiomatization for the Shannon entropy, which will serve as a basis for our axiomatization. A probability measure $p(\{1\}) = x, p(\{2\}) = 1 - x$ on $N = \{1, 2\}$ is denoted by pair $(x, 1 - x)$ and the entropy of it is denoted by $H(x, 1 - x)$ instead of $H((x, 1 - x))$.

(F1) $f(x) := H(x, 1-x)$ is continuous on $0 \leq x \leq 1$, and there exists $x_0 \in [0, 1]$ such that $f(x_0) > 0$.

(F2) For any permutation π on $\{1, \ldots, n\}$,

$$H(p_{\pi(1)}, \ldots, p_{\pi(n)}) = H(p_1, \ldots, p_n).$$

(F3) If $p_n = q + r, q > 0, r > 0$, then

$$H(p_1, \ldots, p_{n-1}, q, r) = H(p_1, \ldots, p_n) + p_n H(q/p_n, r/p_n).$$

Theorem 1 ([4]). *Under the condition* $H(1/2, 1/2) = 1$, *there exists a unique function* $H : \Delta \to [0, 1]$ *satisfying* (F1), (F2) *and* (F3), *and it is given by* H_S.

We introduce further concepts about capacities, which will be useful for stating the axioms of an entropy.

Definition 23 (dual capacity). *Let v be the capacity on $(N, \mathcal{S})$. Then the* dual capacity *of v is defined on $\mathcal{S}^d := \{A \in 2^N \mid A^c \in \mathcal{S}\}$ by $v^d(A) := 1 - v(A^c)$ for any $A \in \mathcal{S}^d$, where $A^c := N \setminus A$.*

Definition 24 (permutation of v). *Let π be a permutation on N. Then the* permutation of v *by π is defined on $\pi(\mathcal{S}) := \{\pi(A) \in 2^N \mid A \in \mathcal{S}\}$ by $\pi \circ v(A) := v(\pi^{-1}(A))$.*

Let us consider a chain of length 2 as a set system, denoted by $\mathbf{2}$ (e.g., $\{\emptyset, \{1\}, \{1, 2\}\}$), and a capacity $v^{\mathbf{2}}$ on it. We denote by the triplet $(0, u, 1)$ the values of $v^{\mathbf{2}}$ along the chain and we suppose $\mathbf{2} := \{\emptyset, \{1\}, \{1, 2\}\}$ unless otherwise noted.

Definition 25 (embedding of $v^{\mathbf{2}}$). *Let v be a capacity on a totally ordered regular set system $(N, \mathcal{S})$, where $\mathcal{S} := \{C_0, \ldots, C_n\}$ such that $C_{i-1} \prec C_i$, $i = 1, \ldots, n$, and let $v^{\mathbf{2}} := (0, u, 1)$ be a capacity on $\mathbf{2}$. Then for $C_k \in \mathcal{S}$, v^{C_k} is called the* embedding *of $v^{\mathbf{2}}$ into v at C_k and defined on the totally ordered regular set system $(N^{C_k}, \mathcal{S}^{C_k})$ by*

$$v^{C_k}(A) := \begin{cases} v(A), & \text{if } A = C_j, j < k, \\ v(C_{k-1}) + u \cdot (v(C_k) - v(C_{k-1})), & \text{if } A = C'_k \\ v(C_{j-1}), & \text{if } A = C'_j, j > k, \end{cases} \quad (22)$$

where $\{i_k\} := C_k \setminus C_{k-1}, i'_k \neq i''_k$, $(N \setminus \{i_k\}) \cap \{i'_k, i''_k\} = \emptyset, N^{C_k} := (N \setminus \{i_k\}) \cup \{i'_k, i''_k\}, C'_k := (C_k \setminus \{i_k\}) \cup \{i'_k\}, C'_j := (C_{j-1} \setminus \{i_k\}) \cup \{i'_k, i''_k\}$ for $j > k$, and $\mathcal{S}^{C_k} := \{C_0, \ldots, C_{k-1}, C'_k, C'_{k+1}, \ldots, C'_{n+1}\}$.

Remark that more precisely, the dual capacity of v is the dual capacity space of the capacity space $(N, \mathcal{S}, v)$ which is defined by $(N, \mathcal{S}, v)^d := (N, \mathcal{S}^d, v^d)$ with $\mathcal{S}^d := \{A^c \in 2^N \mid A \in \mathcal{S}\}$, the permutation of v is the permutation of the capacity space $(N, \mathcal{S}, v)$ which is defined by $(N, \mathcal{S}, v)^\pi := (N, \pi(\mathcal{S}), \pi \circ v)$, and the embedding of $(0, u, 1)$ into v is the embedding of the capacity space $(\{1, 2\}, \mathbf{2}, (0, u, 1))$ into the capacity space $(N, \mathcal{S}, v)$, and it is defined by $(N, \mathcal{S}, v)^{C_k} := (N^{C_k}, \mathcal{S}^{C_k}, v^{C_k})$.

Now we introduce five axioms for Marichal-Roubens's entropy of a capacity.

(A1) (*continuity*) *The function* $f(u) := H(\{1, 2\}, \mathbf{2}, (0, u, 1)) = H(0, u, 1)$ *is continuous on* $0 \leq u \leq 1$, *and there exists* $u_0 \in [0, 1]$ *such that* $f(u_0) > 0$.

(A2) (*dual invariance*) *For any capacity* $(0, u, 1)$ *on* $\mathbf{2}$,

$$H(0, u, 1) = H(0, 1 - u, 1).$$

(A3) (*increase by embedding*) *Let* v *be a capacity on a totally ordered set system* $(N, \mathcal{S})$. *For any* $c_k \in \mathcal{S}$, *for any* $v^\mathbf{2} := (0, u, 1)$, *the entropy of* v^{c_k} *is*

$$\begin{aligned} H(N, \mathcal{S}, v)^{c_k}) &= H(N, \mathcal{S}, v) + (v(c_k) - v(c_{k-1})) \cdot H(\{1, 2\}, \mathbf{2}, v^\mathbf{2}) \\ &= H(v) + (v(c_k) - v(c_{k-1})) \cdot H(v^\mathbf{2}). \end{aligned}$$

(A4) (*convexity*) *Let* $(N, \mathcal{S})$, $(N, \mathcal{S}_1), (N, \mathcal{S}_2)$ *and* $(N, \mathcal{S}_m)$ *be regular set systems satisfying* $\mathcal{M}(\mathcal{S}) = \mathcal{M}(\mathcal{S}_1) \cup \cdots \cup \mathcal{M}(\mathcal{S}_m)$, *and* $\mathcal{M}(\mathcal{S}_i) \cap \mathcal{M}(\mathcal{S}_j) = \emptyset, i \neq j$. *Then there exists an* $\alpha_1, \alpha_2, \ldots, \alpha_m \in]0, 1[, \alpha_1 + \cdots + \alpha_m = 1$ *such that for any capacity* v *on* $(N, \mathcal{S})$,

$$\begin{aligned} H(N, \mathcal{S}, v) &= \alpha_1 H(N, \mathcal{S}_1, v|_{\mathcal{S}_1}) + \cdots + (\alpha_m) H(N, \mathcal{S}_m, v|_{\mathcal{S}_m}) \\ &= \alpha_1 H(v|_{\mathcal{S}_1}) + \cdots + (\alpha_m) H(v|_{\mathcal{S}_m}). \end{aligned}$$

(A5) (*permutation invariance*) *Let* v *be a capacity on* $(N, 2^N)$. *Then for any permutation* π *on* N, *it holds that*

$$H(N, 2^N, v) = H(N, 2^N, v \circ \pi).$$

The following can be shown.

Theorem 2 ([9]). *Under the condition* $H(\{1, 2\}, \mathbf{2}, (0, \frac{1}{2}, 1)) = H(0, \frac{1}{2}, 1) = 1$, *there exists a unique function* $H : \Delta' \to [0, 1]$ *satisfying* (A1), (A2), (A3), (A4) *and* (A5), *and it is given by* H_{MR}.

We discuss in detail our axioms, in the light of Faddeev's axioms.

- *continuity*

By $f(u) = H_{HG}(0, u, 1) = H_{\mathrm{S}}(p^{(0,u,1),\mathcal{C}^\mathbf{2}}) = H_{\mathrm{S}}(u, 1 - u)$, where $\mathcal{C}^\mathbf{2} := (\emptyset, \{1\}, \{1, 2\})$, (A1) corresponds to (F1).

- *dual invariance*

  $H_{\mathrm{MR}}(v)$ is dual invariant, when v is not only a capacity on $\mathbf{2}$ but also on general regular set systems, that is, for any capacity v on a regular set system we have $H_{\mathrm{MR}}(v) = H_{\mathrm{MR}}(v^d)$. On the other hand, the Shannon entropy of the probability measure is also dual invariance, as a matter of fact, the probability measure and its dual measure are identical.

- *increase by embedding*

  Let v be a capacity on a totally ordered set system $\mathcal{S} = \{\emptyset = c_0, c_1, \ldots, c_n = N\}$, where $c_i \subset c_j$ for $i < j$, and consider the embedding into v with $v^{\mathbf{2}} := (0, u, 1)$ at c_k. Then

$$H_{\mathrm{MR}}(v^{c_k}) = H_{\mathrm{S}}(p^{v^{c_k}, \mathcal{C}'}),$$

where $\mathcal{C}' := (c_0, \ldots, c_{k-1}, c_{k'}, c_{k''}, c_{k+1}, \ldots, c_n)$, and by (F3), we have

$$H_{\mathrm{S}}(p^{v^{c_k}, \mathcal{C}'}) = H_{\mathrm{S}}(p^{v, \mathcal{C}}) + (v(c_k) - v(c_{k-1})) \cdot H_{\mathrm{S}}(u, 1-u)$$

which can be rewritten as

$$H_{\mathrm{MR}}(v^{c_k}) = H_{\mathrm{MR}}(v) + (v(c_k) - v(c_{k-1})) \cdot H_{\mathrm{MR}}(v^{\mathbf{2}}).$$

This is exactly (A3).

- *permutation invariance*

  Let $N := (1, 2, 3)$ and $\mathcal{S} := \{\emptyset, \{1\}, \{3\}, \{1, 2\}, \{1, 3\}, \{2, 3\}, N\}$ and let $\pi = \begin{pmatrix} 1 & 2 & 3 \\ 2 & 3 & 1 \end{pmatrix}$. Then, for instance

$$v \circ \pi(\{2, 3\}) = v(\pi^{-1}(\{2, 3\})) = v(\{1, 2\})$$

(cf. Fig. 9).

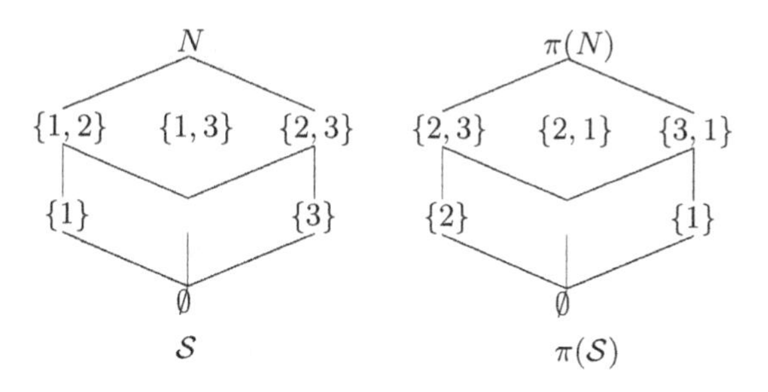

Fig. 9. Permutation of set system

When $\mathcal{S} = 2^N$, all permutations satisfy $\pi(A) \in \mathcal{S}$ for any $A \in \mathcal{S}$. In other words, (A5) can be regarded as a generalization of (F2).

To finish this section, we consider a modification of our axiomatization so as to recover the minimum entropy $H_{\min}$. We modify (A4) as follows:

(A4′) *Let* $(N, \mathcal{S})$, $(N, \mathcal{S}_1)$ *and* $(N, \mathcal{S}_2)$ *be regular set systems satisfying* $\mathcal{M}(\mathcal{S}) = \mathcal{M}(\mathcal{S}_1) \cup \mathcal{M}(\mathcal{S}_2)$. *Then for any capacity* v *on* $\mathcal{S}$,

$$H(v) = \min\left\{H(v|_{\mathcal{S}_1}), H(v|_{\mathcal{S}_2})\right\}.$$

Theorem 3 ([9]). *Under the condition* $H_{\min}(N, \mathbf{2}, (0, \frac{1}{2}, 1)) = H_{\min}(0, \frac{1}{2}, 1) = 1$, *there exists a unique function* $H : \Delta' \to [0,1]$ *satisfying* (A1), (A2), (A3) *and* (A4′), *and it is given by* $H_{\min}$.

6 Conclusions

In this chapter, we discussed entropies of capacities, especially defined by Yager, Marichal-Roubens and Dukhovny.

These entropies, defined for classical capacities, can be generalized for capacities defined on regular set systems and each of them has reasonable properties. And we showed that almost all capacities which appear in applications can be regarded as capacities on regular set systems. In the last section, we showed axiomatizations for Marichal-Roubens's entropy and Dukhobny's entropy, which give validity to them. In this connection, Kojadinovic et al. showed another axiomatization for Marichal-Roubens's entropy [11].

To study the meaning of Yager's entropy more, characterizations by axiomatizations of it are desired.

References

1. Bilbao, J.M.: Cooperative games on combinatorial structures. Kluwer Academic Publishers, Boston (2000)
2. Davey, B.A., Priestley, H.A.: Introduction to lattices and order. Cambridge University Press (1990)
3. Dukhovny, A.: General entropy of general measures. Internat. J. Uncertain. Fuzziness Knowledge-Based Systems 10, 213–225 (2002)
4. Faddeev, D.K.: The notion of entropy of finite probabilistic schemes. Uspekhi Mat. Nauk. 11, 15–19 (1956) (Russian)
5. Faigle, U., Kern, W.: The Shapley value for cooperative games under precedence constraits. Int. J. of Game Theory 21, 249–266 (1992)
6. Grabisch, M., Labreuche, C.: Bi-capacities — Part I: definition, Mobius transform and interaction. Fuzzy Sets and Systems 151, 211–236 (2005)
7. Grabisch, M., Labreuche, C.: Bi-capacities — Part II: the Choquet integral. Fuzzy Sets and Systems 151, 237–259 (2005)
8. Honda, A., Grabisch, M.: Entropy of capacities on lattices. Information Sciences 176, 3472–3489 (2006)
9. Honda, A., Grabisch, M.: An axiomatization of entropy of capacities on set systems. European Journal of Operational Research 190, 526–538 (2008)
10. Hsiao, C.R., Raghavan, T.E.S.: Shapley value for multichoice cooperative games. I. Games and Economic Behavior 5, 240–256 (1993)
11. Kojadinovic, I., Marichal, J.-L., Roubens, M.: An axiomatic approach to the definition of the entropy of a discrete Choquet capacity. Information Sciences 172, 131–153 (2005)

12. Marichal, J.-L., Roubens, M.: Entropy of discrete fuzzy measure. Internat. J. Uncertain. Fuzziness Knowledge-Based Systems 8, 625–640 (2000)
13. Shannon, C.E.A.: Mathematical theory of communication. Bell System Tech. Journ. 27, 374–423 (1948)
14. Shapley, L.S.: A value for n-person games. In: Kuhn, H.W., Tucker, A.W. (eds.) Contributions to the Theory of Games, vol. II, pp. 307–317. Annals of Mathematics Studies 28. Princeton University Press (1953)
15. Sugeno, M.: Fuzzy measures and fuzzy integrals: a survey. In: Gupta, M.M., Saridis, G.N., Gains, B.R. (eds.) Fuzzy Automata and Decision Processes, pp. 89–102. North-Holland, Amsterdam (1977)
16. Yager, R.R.: On the entropy of fuzzy measures. IEEE Transaction on Fuzzy Systems 8, 453–461 (2000)

Integral with Respect to Non-additive Measure in Economics

Hiroyuki Ozaki*

Faculty of Economics, Keio University

Abstract. This chapter surveys the use of non-additive measures in economics, focusing on their use in preference theory. In economics, the risky situation where the probability measure is known and the uncertain situation where even the probability measure is unknown had tended not to be distinguished. This is mainly because of Savage's theorem which states that if an agent complies with some set of behavioral axioms, she may be regarded as trying to maximize the expected utility with respect to *some* probability measure. This is called subjective expected utility (SEU) theory. However, the plausible and robust preference patterns which cannot be explained by SEU is known. The most famous one among them is Ellsberg's paradox. The attempts to resolve these anomalies by using non-additive measures were initiated by D. Schmeidler and I. Gilboa in 1980's. The main purpose of this chapter is to explain their theories, emphasizing representation theorems by means of non-additive measures. We will see that their models nicely resolve Ellsberg's paradox.

1 Introduction

In economics, it is common to distinguish an *uncertain* situation from a *risky* situation at least conceptually. While both seem to be similar in the sense that they represent a situation which is not certain, there is a clear difference between them. The term "risk" used in economics does not necessarily imply the negative aspects such as disaster or danger suggested by that term. For example, the risk in "risk management" in financial technology does not mean only a loss (not to speak of only a gain) but means that both of a loss and a gain may arise. In this example, a risk means the situation where the probability that a loss arises (for instance, a half) and the probability that a gain arises (for instance, a half) are known. That is, economics defines a *risky* situation as the one where, while which event will happen is not exactly known, the probability of the occurrence of each event is known. On the other hand, an *uncertain* situation is defined as the one where even the probability of each event is not known. The uncertainty in this sense is sometimes referred to as *Knightian uncertainty* after an economist who emphasized the difference between these two concepts.

* The author is very grateful to the editors of the book and three anonymous referees for very useful comments. All remaining errors are of course mine.

V. Torra, Y. Narukawa, and M. Sugeno (eds.), *Non-Additive Measures*, Studies in Fuzziness and Soft Computing 310,
DOI: 10.1007/978-3-319-03155-2_5, © Springer International Publishing Switzerland 2014

Despite of such a clear difference between them, it has been widely customary not to distinguish these two concepts in economic analyses. It is so-called Savage's Theorem proven by a statistician named Leonard J. Savage in 1954 that justifies this custom. According to this theorem, under some conditions, even if there does not exist an *objective* probability (such as a half and a half of a coin flip), we may think that people behave as if they know the probability (which is called a *subjective* probability). That is, an uncertain situation is reduced to a risky situation.

Although situations where the conditions of Savage's Theorem do not hold are known for a long time (for example, Ellsberg's Paradox in 1961), it had been quite often regarded as exceptional. However, in recent years (about for the last two decades), some approaches have been developed in which uncertainty is directly analyzed without reducing it to a risk. By this, economic phenomena, which cannot be explained within the traditional framework by Savage, has been made clearer.

The objective of this chapter is to elucidate utility-theoretic approaches toward uncertainty developed by Schmeidler, Gilboa, and so on which utilize non-additive measures, in particular focusing on their axiomatic foundations. First, in order to clarify the approaches by Schmeidler and Gilboa, we explain subjective expected utility theories by Savage, Anscombe and Aumann, and so on (and objective expected utility theories by von-Neumann and Morgenstern, and so on as their bases) including their proofs. Next, we introduce Ellsberg's Paradox. This example directly shows the limitation of representation of preferences by probability measures which is a common characteristic of above theories. We then explain representation theorems by non-additive measures which "resolve" this paradox. In particular, we aim to clarify the relationship between the axioms and representation, and we emphasize the mathematical structure of the representation and the proofs of representation theorems. (The proofs are adopted from Fishburn (1970), Kreps (1988) and the original papers and are suitably arranged.) We also briefly mention an economic application of the theory. The chapter is concluded with an appendix that summarizes some results on non-additive measures. The readers are referred to the book by Gilboa (2009) to see more on related topics to the ones covered in this chapter.

2 Some Definitions on Probability Measures

We call a family of subsets $\mathcal{A}$ of a set X an *algebra* if it satisfies the three conditions: (1) $\phi \in \mathcal{A}$, (2)[1] $A \in \mathcal{A} \Rightarrow A^c \in \mathcal{A}$ and (3) $A, B \in \mathcal{A} \Rightarrow A \cup B \in \mathcal{A}$, and call a pair of a set and an algebra defined on that set, $(X, \mathcal{A})$, a *measurable space*. The family of subsets of a given set X consisting of all its subsets is called a *power set* and denoted by 2^X. Clearly, $(X, 2^X)$ is a measurable space. In this chapter, we do *not* consider a σ-algebra.[2] Given a measurable space $(X, \mathcal{A})$, a set

[1] Here, A^c denotes the complement of A in X.

[2] An algebra which is closed with respect to a union of countably many sets is called an σ-*algebra*.

function $p : \mathcal{A} \to [0, +\infty]$ which satisfies the following two conditions is called a *finitely-additive measure* or a *charge*:

$$p(\phi) = 0 \tag{1}$$

$$(\forall A, B \in \mathcal{A}) \quad A \cap B = \phi \Rightarrow p(A \cup B) = p(A) + p(B). \tag{2}$$

Condition (2) is called a *finite additivity*. It immediately follows that a finitely-additive measure p is *monotonic* in the sense that[3]:

$$A \subseteq B \Rightarrow p(A) \leq p(B).$$

A finitely-additive measure p which satisfies $p(X) = 1$ is called a finitely-additive *probability* measure. In what follows, we call a finitely-additive probability measure simply a *probability measure*. Note that a probability measure does *not* necessarily satisfy the σ-additivity.[4]

A probability measure p is said to be *simple* if the set of $x \in X$ such that $p(x) \neq 0$ is a finite set and if $\sum_{x \in X} p(x) = 1$ holds. In particular, we write the simple probability measure such that $p(x) = 1$ for some $x \in X$ as δ_x. Also, a probability measure p on a measurable space $(X, \mathcal{A})$ is said to be *convex-ranged* or *strongly nonatomic* if it satisfies the next condition:[5]

$$(\forall A \in \mathcal{A})(\forall r \in [0, p(A)])(\exists B \in \mathcal{A}) \quad B \subseteq A \text{ and } p(B) = r$$

3 Expected Utility Theory à la von-Neumann-Morgenstern

3.1 Preference Order

We consider a set X to be a set of rewards, prizes or consequences and call it the *set of outcomes*. Let $\mathcal{A}$ be an algebra on X. We denote by $P(X)$ the set of all probability measures on the measurable space $(X, \mathcal{A})$ and write generic elements of $P(X)$ (*i.e.*, probability measures) as $p, q, r, \ldots$. We call any subset $\succ$ of $P(X) \times P(X)$ a *binary relation* and write as $p \succ q$ when $(p, q) \in \succ$.

[3] Proof: For any A and B such that $A \subseteq B$, it holds that $B = A \cup (B \backslash A)$ and $A \cap (B \backslash A) = \phi$. This implies that $p(B) = p(A) + p(B \backslash A) \geq p(A)$. Here, the equality follows from the finite additivity and the inequality follows from the fact that p takes on nonnegative values.

[4] A finitely-additive measure p is said to satisfy *σ-additivity* if the similar condition to (2) holds for countably many sets.

[5] A probability measure p on a measurable space $(X, \mathcal{A})$ is said to be *nonatomic* if

$$(\forall A \in \mathcal{A}) \quad p(A) > 0 \Rightarrow (\exists B \in \mathcal{A}) \, B \subseteq A \text{ and } p(B) \in (0, p(A)).$$

While if a probability measure p is convex-ranged, it is nonatomic, the converse does not necessarily hold. If an algebra $\mathcal{A}$ happens to be a σ-algebra and if p happens to be σ-additive, both are equivalent.

A binary relation $\succ$ is said to be *asymmetric* if

$$(\forall p, q \in P(X)) \quad p \succ q \Rightarrow q \not\succ p,$$

where $q \not\succ p$ means that $(q, p) \notin \succ$, and it is said to be *negatively transitive* if

$$(\forall p, q, r \in P(X)) \quad p \not\succ q \text{ and } q \not\succ r \Rightarrow p \not\succ r.$$

A binary relation $\succ$ is called a *preference order* or a *preference relation* when it is asymmetric and negatively transitive. Given a preference order $\succ$, the binary relation $\succeq$ is defined by

$$(\forall p, q \in P(X)) \quad p \succeq q \Leftrightarrow q \not\succ p$$

and the binary relation $\sim$ is defined by

$$(\forall p, q \in P(X)) \quad p \sim q \Leftrightarrow p \not\succ q \text{ and } q \not\succ p.$$

Then, $\succeq$ turns out to be transitive and complete, where $\succeq$ is *transitive* by definition if

$$(\forall p, q, r \in P(X)) \quad p \succeq q \text{ and } q \succeq r \Rightarrow p \succeq r$$

and it is *complete* by definition if

$$(\forall p, q \in P(X)) \quad p \succeq q \text{ or } q \succeq p \text{ or both.}$$

A binary relation is called a *weak order* if it is transitive and complete. We may define a weak order first and then derive a binary relation so that the latter should be a preference order. Also, a binary relation $\sim$ is called an *indifference relation*.

Let $u : P(X) \to \mathbb{R}$ be a real-valued function defined on $P(X)$. A function u is said to *represent* a preference order $\succ$ if it holds that

$$(\forall p, q \in P(X)) \quad p \succ q \Leftrightarrow u(p) > u(q)$$

and it is said to be an *affine function* if it holds that

$$(\forall p, q \in P(X))(\lambda \in [0,1]) \quad u(\lambda p + (1 - \lambda)q) = \lambda u(p) + (1 - \lambda)u(q).$$

In the above, similar definitions apply when $P(X)$ is replaced by its arbitrary convex subset.

3.2 vNM Axioms and Representation Theorem

Consider the following three axioms with respect to a binary relation $\succ$ defined on a convex subset of $P(X)$. Here, p, q, r are arbitrary elements of that set and λ is an arbitrary real number such that $\lambda \in (0, 1)$.

A1 (Ordering) $\succ$ is a preference order

A2 (Independence) $p \succ q \;\Rightarrow\; \lambda p + (1-\lambda)r \succ \lambda q + (1-\lambda)r$

A3 (Continuity) $p \succ q$ and $q \succ r$
$\Rightarrow\; (\exists \alpha, \beta \in (0,1))\;\; \alpha p + (1-\alpha)r \succ q$ and
$q \succ \beta p + (1-\beta)r$

Then, the following theorem holds.

Theorem 1 (von-Neumann-Morgenstern, 1944). *Let $\succ$ be a binary relation defined on an arbitrary convex subset $\mathcal{P}$ of $P(X)$. Then, $\succ$ satisfies Axioms A1, A2 and A3 if and only if there exists an affine function u which represents $\succ$. Furthermore, if a function u' also represents $\succ$, then there exist real numbers $a > 0$ and b such that $u' = au + b$.*

To prove the theorem we first prove the next lemma.

Lemma 1. *If a binary relation $\succ$ on $\mathcal{P}$ satisfies Axioms A1, A2 and A3, then the following holds:*
(a) If $p \succ q$ and $0 \le a < b \le 1$, then $bp + (1-b)q \succ ap + (1-a)q$.
(b) If $p \succeq q \succeq r$ and $p \succ r$, then there exists a unique $a^* \in [0,1]$ such that $q \sim a^*p + (1-a^*)r$.
(c) If $p \sim q$ and $a \in [0,1]$, then for any $r \in \mathcal{P}$, it holds that $ap + (1-a)r \sim aq + (1-a)r$.

Proof of (a). If $a = 0$, then the claim follows immediately from A2. Hence, assume that $a > 0$. Define r by $r := bp + (1-b)q$. Then, A2 implies $r \succ q$, and hence, it follows that

$$\begin{aligned} r &= (1-(a/b))r + (a/b)r \succ (1-(a/b))q + (a/b)r \\ &= (1-(a/b))q + (a/b)(bp + (1-b)q) = ap + (1-a)q\,, \end{aligned}$$

which completes the proof. Proof of (b). If a^* exists, it must be unique by (a). Therefore, it suffices to show the existence of a^*. Assume that $p \succ q \succ r$. Otherwise, the existence of a^* is trivial. Define a real number a^* by

$$a^* = \sup\{\, a \in [0,1] \mid q \succeq ap + (1-a)r \,\}\,.$$

First, assume that $a^*p + (1-a^*)r \succ q \succ r$. Then, A3 implies that there exists $b \in (0,1)$ such that $b(a^*p + (1-a^*)r) + (1-b)r = ba^*p + (1-ba^*)r \succ q$. Furthermore, $ba^* < a^*$ holds since $a^* \neq 0$. This and the definition of a^* imply the existence a' such that $ba^* < a' < a^*$ and $q \succeq a'p + (1-a')r$. Then, (a) implies $q \succ ba^*p + (1-ba^*)r$, which is a contradiction. Similarly, assuming $p \succ q \succ a^*p + (1-a^*)r$ leads to a contradiction. Therefore, (b) holds. The proof of (c) is omitted. ∎

Proof (Sketch). We now prove the theorem. When all probability measures are indifferent to each other, we may take any constant function as u. Hence, we assume that there exists a pair of probability measures p and q such that $p \succ q$ and fix them. For any r such that $p \succeq r \succeq q$, we define $u(r)$ by a real number which satisfies

$$u(r)p + (1 - u(r))q \sim r\,.$$

By (b) of Lemma 1, such a $u(r)$ is uniquely determined. Then, for any r and r' such that $p \succeq r \succeq q$ and $p \succeq r' \succeq q$, it holds that

$$r \succ r' \Leftrightarrow u(r)p + (1 - u(r))q \succ u(r')p + (1 - u(r'))q \Leftrightarrow u(r) > u(r')$$

where the second equivalence follows from (a) of Lemma 1. Furthermore, for any such r and r' and for any $\lambda \in [0, 1]$, it holds that

$$\begin{aligned}\lambda r + (1-\lambda)r' &\sim \lambda(u(r)p + (1 - u(r))q) + (1-\lambda)(u(r')p + (1 - u(r'))q)\\ &= (\lambda u(r) + (1-\lambda)u(r'))p + (1 - (\lambda u(r) + (1-\lambda)u(r')))q\,.\end{aligned}$$

Here, the equivalence follows from (c) of Lemma 1. This and the definition of u imply

$$u(\lambda r + (1-\lambda)r') = \lambda u(r) + (1-\lambda)u(r')\,.$$

The function u thus defined preserves its properties through a positive affine transformation. Therefore, we may assume without loss of generality that $u(p) = 1$ and $u(q) = 0$. Further, for any r such that $r \succ p$, find an a which satisfies $ar + (1-a)q \sim p$ (such an a can be found uniquely) and define $u(r) := 1/a$. Also, for any r such that $q \succ r$, find an a which satisfies $ap + (1-a)r \sim q$ (such an a can be found uniquely) and define $u(r) := -a/(1-a)$. The function u thus defined can be easily verified to be an affine function which represents $\succ$ on $\mathcal{P}$ and the proof is complete. ■

We say that u is *unique up to* a positive affine transformation when it is unique in the sense stated in Theorem 1. A function u in the theorem is called von-Neumann-Morgenstern's *utility index*. When a utility index u exists, define $u : X \to \mathbb{R}$ by $(\forall x \in X)\;\; u(x) = u(\delta_x)$. Then the affinity of u implies that for any simple probability measure p, it holds that

$$u(p) = \int_X u(x)\, dp(x) = \sum_{x \in X} u(x)p(x)\,.$$

4 Representation Theorem on Mixture Space à la Herstein-Milnor

A set Φ is called a *mixture space* if there exists a function $h : [0,1] \times \Phi \times \Phi \to \Phi$ which satisfies the following three conditions:[6]

[6] In this section, ϕ denotes an arbitrary element of Φ, not an empty set.

M1 $h_1(\phi, \rho) = \phi$

M2 $h_a(\phi, \rho) = h_{1-a}(\rho, \phi)$

M3 $h_a(h_b(\phi, \rho), \rho) = h_{ab}(\phi, \rho)$

Here, the first argument is denoted by a subscript and $h_a(\phi, \rho)$ may be suggestively written as $a\phi + (1-a)\rho$. However, a mixture space is purely an abstract space and "+" need not be an addition (although in many examples, "+" resolves into addition of real numbers). For instance, the set of probability measures considered in the previous section is an example of a mixture space. Consider the following three axioms with respect to a binary relation $\succ$ defined on a mixture space:

B1 (Ordering) $\succ$ is a preference order

B2 (Independence) $\phi \succ \rho \;\Rightarrow\; (\forall a \in (0,1])(\forall \mu)\;\; h_a(\phi, \mu) \succ h_a(\rho, \mu)$

B3 (Continuity) $\phi \succ \rho$ and $\rho \succ \mu$
$\Rightarrow\; (\exists a, b \in (0,1))\;\; h_a(\phi, \mu) \succ \rho$ and
$\rho \succ h_b(\phi, \mu)$

Then, the following theorem holds.

Theorem 2 (Herstein-Milnor, 1953). *A binary relation $\succ$ defined on a mixture space Φ satisfies Axioms B1, B2 and B3 if and only if there exists a function $F : \Phi \to \mathbb{R}$ which satisfies*

Representation $\phi \succ \rho \;\Rightarrow\; F(\phi) > F(\rho)$ *and*

Affinity $(\forall a, \phi, \rho)\quad F(h_a(\phi, \rho)) = aF(\phi) + (1-a)F(\rho)$.

Furthermore, F is unique up to a positive affine transformation.

The proof of this theorem can be done exactly similarly to the case of von-Neumann-Morgenstern's Theorem (Theorem 1). This is because the proof of that theorem uses only the properties of a mixture space. (Of course, an arbitrary convex subset of $P(X)$ is a mixture space.)

5 Subjective Expected Utility Theory à la Savage

5.1 Subjective Probability Theorem

Let S be a set of states which may happen and we call S a *state space* or a set of *states of the world*. Also, let $\mathcal{E}$ be an algebra consisting of subsets of S. An element of $\mathcal{E}$ is called an *event*. In what follows, we set $\mathcal{E} = 2^S$. A binary relation $\succ_\ell$ on the algebra $\mathcal{E}$ is said to be a *qualitative probability* if it satisfies the following four conditions:

QP1 $\succ_\ell$ is (formally) a preference order

QP2 $(\forall E \in \mathcal{E})\ E \succeq_\ell \phi$

QP3 $S \succ_\ell \phi$

QP4 $(\forall E, F, G \in \mathcal{E})$
$E \cap G = F \cap G = \phi \Rightarrow [E \succ_\ell F \Leftrightarrow E \cup G \succ_\ell F \cup G]$

If a binary relation $\succ_\ell$ is a qualitative probability, it follows that

$$(\forall A, B) \quad A \subseteq B \Leftrightarrow B \succeq_\ell A\,.$$

Also, it is immediate that $\succ_\ell$ must be a qualitative probability if it is representable by a probability measure. To show its converse, the next condition is necessary.

QP5 $F \succ_\ell G \Rightarrow (\exists \langle E_i \rangle_{i=1}^n \subseteq \mathcal{E})(\forall k = 1, \ldots, n)\ F \succ_\ell G \cup E_k$

Savage (1954) proved the next theorem.

Theorem 3 (Subjective Probability Theorem). *A binary relation $\succ_\ell$ is a qualitative probability which satisfies Condition QP5 if and only if there exists a unique convex-ranged probability measure on the measurable space $(S, \mathcal{E})$ which represents $\succ_\ell$.*

5.2 Act

Let X be a set of outcomes. A function from S into X is called a *Savage act*, or more simply, an *act*. An act f is said to be *simple* if the image of S by f, $f(S) := \{x \in X \mid (\exists s \in S)\ f(s) = x\}$, is a finite set and the set of all simple act is denoted by F_0. In what follows, when we say an act, it means a simple act if otherwise stated. Also, denote by $P_0(X)$ the set of all simple probability measures on X. Assume that an agent's preference is given by a binary relation $\succ$ on F_0 and binary relations $\succeq$ and $\sim$ are derived from $\succ$ in the way already mentioned. An event $E \in \mathcal{E}$ is said to be *null* with respect to $\succ$ if any pair of acts which differ only on E is indifferent to each other. We induce a binary relation on X from a binary relation $\succ$ on F_0 and denote it by the same symbol $\succ$ as follows (using the same symbol will not make any confusion):

$$(\forall x, y \in X) \quad x \succ y \Leftrightarrow f \succ g \text{ where } (\forall s)\ f(s) = x \text{ and } g(s) = y\,.$$

In what follows, we use a notation such as

$$f = \begin{bmatrix} x \text{ on } A \\ g \text{ on } A^c \end{bmatrix}$$

to denote an act f such that f always takes on an outcome $x \in X$ on an event A (*i.e.*, f coincides with a constant act x on A) and coincides with an act $g \in L_0$ on A^c, the complement of A in S.

5.3 Savage's Axioms and Representation Theorem by Subjective Expected Utility

Consider the following axioms:

P1. Ordering $\succ$ is a preference order

P2. Sure-Thing Principle For any act f, f', g, g' and for any event A,

$$\begin{bmatrix} f \text{ on } A \\ g \text{ on } A^c \end{bmatrix} \succ \begin{bmatrix} f' \text{ on } A \\ g \text{ on } A^c \end{bmatrix} \Leftrightarrow \begin{bmatrix} f \text{ on } A \\ g' \text{ on } A^c \end{bmatrix} \succ \begin{bmatrix} f' \text{ on } A \\ g' \text{ on } A^c \end{bmatrix}$$

P3. Eventwise Monotonicity For any outcome x and y, for any non-null event E and for any act f,

$$\begin{bmatrix} x \text{ on } E \\ f \text{ on } E^c \end{bmatrix} \succ \begin{bmatrix} y \text{ on } E \\ f \text{ on } E^c \end{bmatrix} \Leftrightarrow x \succ y$$

P4. Weak Comparative Probability For any event A and B and for any outcome x, x', y, y' such that $x \succ x'$ and $y \succ y'$,

$$\begin{bmatrix} x \text{ on } A \\ x' \text{ on } A^c \end{bmatrix} \succ \begin{bmatrix} x \text{ on } B \\ x' \text{ on } B^c \end{bmatrix} \Leftrightarrow \begin{bmatrix} y \text{ on } A \\ y' \text{ on } A^c \end{bmatrix} \succ \begin{bmatrix} y \text{ on } B \\ y' \text{ on } B^c \end{bmatrix}$$

P5. Nondegeneracy There exists a pair of outcomes (x, x') such that $x \succ x'$

P6. Small Event Continuity For any outcome x and for any act f and g such that $f \succ g$, there exists a finite partition, $\langle E_i \rangle_{i=1}^n$, of S such that

$$(\forall i, j \in \{1, 2, \ldots, n\}) \quad f \succ \begin{bmatrix} x \text{ on } E_i \\ g \text{ on } (E_i)^c \end{bmatrix} \text{ and } \begin{bmatrix} x \text{ on } E_j \\ f \text{ on } (E_j)^c \end{bmatrix} \succ g$$

Then, the next theorem holds:

Theorem 4 (Savage, 1954). *A binary relation $\succ$ on F_0 satisfies P1, P2, P3, P4, P5 and P6 if and only if there exist a unique convex-ranged probability measure μ on $(S, \mathcal{E})$ and a real-valued function u on X which is unique up to a positive affine transformation such that*

$$f \succ g \Leftrightarrow \int_S u(f(s))\, d\mu(s) > \int_S u(g(s))\, d\mu(s)\,.$$

Proof (Sketch) For any pair of events A and B, define $A \succ_\ell B$ when it holds that for some outcomes x and x' such that $x \succ x'$,

$$\begin{bmatrix} x \text{ on } A \\ x' \text{ on } A^c \end{bmatrix} \succ \begin{bmatrix} x \text{ on } B \\ x' \text{ on } B^c \end{bmatrix}.$$

P4 implies that $\succ_\ell$ is a preference order: *i.e.*, $\succ_\ell$ does not depend on the choice of x. Also, it follows from P1, P2, P3, P4 and P5 that $\succ_\ell$ turns out to be a qualitative probability. Furthermore, it can be verified that $\succ_\ell$ satisfies QP5 by P6. By Subjective Probability Theorem (Theorem 3), there exists a unique convex-ranged probability measure μ which represents $\succ_\ell$.

From a simple act $f \in F_0$ and the probability measure μ derived in the previous paragraph, define a simple probability measure p_f on X by

$$(\forall x \in X) \quad p_f(x) = \mu(f^{-1}(\{x\}))\,.$$

Denote the set of all simple probability measures on X by $P_0(X)$. For any $p \in P_0(X)$, there exists some $f \in F_0$ such that $p = p_f$. Then, it can be shown that

$$(\forall f, g \in F_0) \quad p_f = p_g \Rightarrow f \sim g\,.$$

From this and the remark just mentioned, we can define a preference relation $\succ_P$ on $P_0(X)$ by

$$(\forall p, q \in P_0(X)) \quad p \succ_P q \Leftrightarrow p = p_f,\ q = p_g \ \text{ and } \ f \succ g$$

Then, since it can be shown that $\succ_P$ satisfies Axioms A1, A2 and A3 of Theorem 1 (or Axioms B1, B2 and B3 of Theorem 2), Theorem 1 (or Theorem 2) implies that there exists a function $u : X \to \mathbb{R}$ which is unique up to a positive affine transformation such that

$$\begin{aligned} f \succ g &\Leftrightarrow p_f \succ_P p_g \\ &\Leftrightarrow \int_X u(x)\, dp_f(x) > \int_X u(x)\, dp_g(x) \\ &\Leftrightarrow \int_S u(f(s))\, d\mu(s) > \int_S u(g(s))\, d\mu(s)\,. \end{aligned}$$

The proof of the necessity of the axioms is omitted. ■

6 Probabilistic Sophistication à la Machina-Schmeidler

Given a simple act f and a probability measure μ on $(S, \mathcal{E})$, we denote by $p_{f,\mu}$ or simply by p_f when no confusion arise with respect to μ the element of $P_0(X)$ defined by $(\forall x \in X)$ $\mu(f^{-1}(\{x\}))$. Also, for any outcome $x \in X$, let δ_x be the element of $P_0(X)$ such that $\delta_x(x) = 1$. Suppose that a binary relation $\succ$ is defined on the set of acts, F_0. For $p, q \in P_0(X)$, we say that p *stochastically dominates* q if

$$(\forall x \in X) \qquad \sum_{\{i \mid x_i \preceq x\}} p_i \le \sum_{\{j \mid y_j \preceq x\}} q_j\,.$$

Here, we write as $p = (x_1, p_1; \ldots; x_m, p_m)$ and $q = (y_1, q_1; \ldots; y_n, q_n)$ and mean that an outcome x_1 occurs with a probability p_1 under p, for example. If a strict inequality holds for some outcome x, we say that p *strictly* stochastically dominates q.

Suppose that a binary relation $\succ_P$ is defined on $P_0(X)$. If for any $p, q, r \in P_0(X)$, $\{\lambda \in [0,1] \mid \lambda p + (1-\lambda)q \succeq_P r\}$ and $\{\lambda \in [0,1] \mid \lambda p + (1-\lambda)q \preceq_P r\}$ are closed sets, then a binary relation $\succ_P$ is said to be *mixture-continuous*. If a binary relation is mixture-continuous, it turns out to be continuous in the sense of Axiom A3. (The converse does not hold.) Also, a binary relation $\succ_P$ is said to be *monotonic* if $p \succeq_P (\succ_P)\, q$ whenever p (strictly) stochastically dominates q. Similarly, if for any $p, q, r \in P_0(X)$, $\{\lambda \in [0,1] \mid V(\lambda p + (1-\lambda)q) \geq V(r)\}$ and $\{\lambda \in [0,1] \mid V(\lambda p + (1-\lambda)q) \leq V(r)\}$ are closed sets, then a function $V : P_0(X) \to \mathbb{R}$ is said to be *mixture-continuous*. Also, a function $V : P_0(X) \to \mathbb{R}$ is said to be *monotonic* if $V(p) \geq (>)\, V(q)$ whenever p (strictly) stochastically dominates q.

Consider the following axioms on a binary relation $\succ$ defined on F_0:

P4*Strong Comparative Probability For any mutually disjoint event A and B, for any outcomes x, x', y, y' such that $x \succ x'$ and $y \succ y'$, and for any act g and h,

$$\begin{bmatrix} x \text{ on } A \\ x' \text{ on } B \\ g \text{ on } (A \cup B)^c \end{bmatrix} \succ \begin{bmatrix} x' \text{ on } A \\ x \text{ on } B \\ g \text{ on } (A \cup B)^c \end{bmatrix}$$

$$\Leftrightarrow \begin{bmatrix} y \text{ on } A \\ y' \text{ on } B \\ h \text{ on } (A \cup B)^c \end{bmatrix} \succ \begin{bmatrix} y' \text{ on } A \\ y \text{ on } B \\ h \text{ on } (A \cup B)^c \end{bmatrix}$$

Axiom P4* implies Axiom P4. Axiom P4* and Axiom P2 are mutually independent but all Savage's axioms imply P4*. For any mutually disjoint event A and B, we write as $A \succ_\ell B$ if there exist a pair of outcomes (x, x') such that $x \succ x'$ and an act g such that

$$\begin{bmatrix} x \text{ on } A \\ x' \text{ on } B \\ g \text{ on } (A \cup B)^c \end{bmatrix} \succ \begin{bmatrix} x' \text{ on } A \\ x \text{ on } B \\ g \text{ on } (A \cup B)^c \end{bmatrix}.$$

Because of Axiom P4*, $\succ_\ell$ does not depend on the choice of x, x', g. Furthermore, if we define $A \succ_\ell B \Leftrightarrow A \backslash B \succ_\ell B \backslash A$ for not necessarily disjoint events A and B, it follows from P4* that $\succ_\ell$ is a preference relation on $(S, \mathcal{E})$. Then, the next theorem holds.

Theorem 5 (Machina and Schmeidler, 1992). *A binary relation $\succ$ on F_0 satisfies P1, P3, P4*, P5 and P6 if and only if there exist a convex-ranged probability measure μ on $(S, \mathcal{E})$ which represents $\succ_\ell$ and a monotonic mixture-continuous function $V : P_0(X) \to \mathbb{R}$ such that $f \succ g \Leftrightarrow V(p_f) > V(p_g)$.*

Proof (Sketch) There exists a pair of outcomes (x, x') such that $x \succ x'$ by Axiom P5. Then, let $X^* := \{x, x'\}$. Also, let F_0^* be the subset of F_0 consisting of all X^*-valued acts and write the restriction of $\succ$ on F_0^* as $\succ^*$. Then, it follows that Axioms P1, P3, P4, P5 and P6 hold with respect to $(S, \mathcal{E}, X^*, F_0^*, \succ^*)$.

Furthermore, it turns out that P4* and P2 are equivalent under this setting, and hence, Savage's Theorem (Theorem 4) implies that there exists a convex-ranged probability measure μ on $(S, \mathcal{E})$ such that for any act $f, g \in F_0^*$, $f \succ g \Leftrightarrow \mu(f^{-1}(\{x\})) \geq \mu(g^{-1}(\{x\}))$ holds. On the other hand, it holds that

$$\begin{aligned} A \succ_\ell B &\Leftrightarrow A \backslash B \succ_\ell B \backslash A \\ &\Leftrightarrow \begin{bmatrix} x \text{ on } A\backslash B \\ x' \text{ on } B\backslash A \\ x' \text{ on } (A \triangle B)^c \end{bmatrix} \succ \begin{bmatrix} x' \text{ on } A\backslash B \\ x \text{ on } B\backslash A \\ x' \text{ on } (A \triangle B)^c \end{bmatrix} \\ &\Leftrightarrow \mu(A\backslash B) > \mu(B\backslash A) \\ &\Leftrightarrow \mu(A) > \mu(B) \end{aligned}$$

and it has been shown that μ represents $\succ_\ell$.

Suppose that $\mu(A) = \mu(B)$ holds for mutually disjoint events A and B. For any outcome y, y' and for any act h, defined two act f and g by

$$f = \begin{bmatrix} y \text{ on } A \\ y' \text{ on } B \\ h \text{ on } (A \cup B)^c \end{bmatrix} \quad \text{and} \quad g = \begin{bmatrix} y' \text{ on } A \\ y \text{ on } B \\ h \text{ on } (A \cup B)^c \end{bmatrix}.$$

Then, since both $A \succeq_\ell B$ and $B \succeq_\ell A$ hold, it follows that $f \sim g$. This means that two acts which are constructed by exchanging outcomes they assume over mutually disjoint events having the same probability are indifferent. By this fact and by applying an appropriately constructed sequence of acts, it can be proved that two acts f and g are indifferent if $p_{f,\mu} = p_{g,\mu}$. From this, it turns out that we may define a binary relation $\succ_P$ on $P_0(X)$ by $p_f \succ_P p_g \Leftrightarrow f \succ g$ as in the proof of Savage's theorem (Theorem 4).

It can be proved that the binary relation $\succ_P$ defined in the previous paragraph satisfies the monotonicity and the mixture-continuity. For $x^*, x_* \in X$ such that $x^* \succ x_*$, define the subset of $P_0(X)$ by

$$\{ p \in P_0(X) \,|\, \delta_{x^*} \succeq_P p \succeq_P \delta_{x_*} \}. \tag{3}$$

Then, it turns out that for any element p of this subset, there exists λ_p which satisfies

$$p \sim_P \lambda_p \delta_{x^*} + (1 - \lambda_p)\delta_{x_*}$$

by the mixture-continuity (and hence by the continuity) of $\succ_P$. (See also the proof of von-Neumann-Morgenstern's theorem.) Furthermore, by the monotonicity of $\succ_P$, λ_p is determined uniquely. Define $V(p) := \lambda_p$. Then, for any element p, q of the set (3) such that $p \succeq_P (\succ_P)\, q$, it holds that

$$V(p)\delta_{x^*} + (1 - V(p))\delta_{x_*} \sim_P p \succeq_P (\succ_P)\, q \sim_P V(q)\delta_{x^*} + (1 - V(q))\delta_{x_*}.$$

From this and the monotonicity, it follows that $V(p) \geq (>)\, V(q)$. That is, the function V represents $\succ_P$ on the set (3). Furthermore, it can be shown that V can be extended so that it should represent $\succ_P$ on $P_0(X)$. The sufficiency of the axioms follows from this fact and the result in the previous paragraph. The proofs of other claims are omitted. ∎

The property that $p_f = p_g \Rightarrow f \sim g$, which is shown in the theorem, is called *probabilistic sophistication.*

7 Subjective Expected Utility Theory à la Anscombe-Aumann

7.1 Lottery Act

In this section, we assume that the state space is given by a measurable space (S, Σ) and that an outcome space is given by a mixture space Y. For example, let X be a space of prizes and let Y be the space of all simple probability measures on X. Then, Y is a mixture space. A function from S into Y is called *Anscombe-Aumann (A-A) act* or *lottery act*. A lottery act which is Σ-measurable and whose range is a finite set is called *simple* lottery act. The set of all simple lottery acts is denoted by L_0. Furthermore, the set of simple lottery acts whose ranges are singletons is denoted by L_c. Suppose that a binary relation $\succ$ is defined on the set L_0. We induce the binary relation on Y from a binary relation $\succ$ on L_0 as follows and denote it by the same symbol $\succ$; *i.e.*,

$$(\forall y, y' \in Y) \quad y \succ y' \ \Leftrightarrow \ f \succ g \ \text{ where } \ (\forall s)\ f(s) = y \text{ and } g(s) = y' .$$

Differently from the framework of Savage, in the one of Anscombe and Aumann, we can construct a mixture of two lottery acts. Let f and g be two simple lottery acts. Also, let h be a mixture function defined on the mixture space Y. Then, use h to define the mixture of f and g by $s \mapsto h_a(f(s), g(s))$. Then, it can be easily seen that the set L_0 becomes a mixture space by the operation thus defined. We write the mixture of f and g by $af + (1 - a)g$.

7.2 Anscombe-Aumann's Axioms and Representation Theorem

We consider some axioms on a binary relation $\succ$ defined on L_0. Here, f, g, h are any element of L_0 and λ is any real number such that $\lambda \in (0, 1]$.

AA1 (Ordering) $\succ$ is a preference order on L_0

AA2 (Independence) $f \succ g \ \Rightarrow \ \lambda f + (1 - \lambda)h \succ \lambda g + (1 - \lambda)h$

AA3 (Continuity) $f \succ g$ and $g \succ h$
$\Rightarrow \ (\exists \alpha, \beta \in (0, 1))\ \alpha f + (1 - \alpha)h \succ g$ and
$g \succ \beta f + (1 - \beta)h$

AA4 (Monotonicity) $(\forall f, g \in L_0)\ [(\forall s \in S)\ f(s) \succeq g(s)] \ \Rightarrow \ f \succeq g$

AA5 (Nondegeneracy) $(\exists f, g \in L_0)\ f \succ g$

The next theorem holds.

Theorem 6 (Anscombe-Aumann, 1963). *A binary relation $\succ$ defined on the set L_0 satisfies AA1, AA2, AA3, AA4 and AA5 if and only if there exist a unique probability measure μ on (S, Σ) and an affine function u on Y which is unique up to a positive affine transformation such that*

$$f \succ g \Leftrightarrow \int_S u(f(s))\, d\mu(s) > \int_S u(g(s))\, d\mu(s)\,.$$

Proof Since the set L_0 is a mixture space and a binary relation $\succ$ satisfies Ordering, Independence and Continuity by the assumption, there exists an affine function on L_0 which represents $\succ$ by Theorem 2. We denote this function by J. Further, define the affine function u on Y by $(\forall y \in Y)\ u(y) = J(y)$, where y in the right-hand side is understood to be a constant act which always takes on y. Monotonicity and Nondegeneracy imply the existence of $y^*, y_* \in Y$ such that $y^* \succ y_*$. Hence, by applying an appropriate affine transformation to J, we can normalize u so that $u(y^*) = 1$ and $u(y_*) = 0$.

Let $K := u(Y)$ and let $B_0(K)$ be the set of Σ-measurable functions defined on S whose range is a finite subset of K. Define a function $U : L_0 \to B_0(K)$ by

$$(\forall f)(\forall s) \quad U(f)(s) = u(f(s))\,.$$

Then, U is surjective and Monotonicity implies that $U(f) = U(g) \Rightarrow f \sim g$. Furthermore, the affinity of u implies $(\forall \alpha \in [0,1])\ U(\alpha f + (1-\alpha)g) = \alpha U(f) + (1-\alpha)U(g)$. Now, define a functional I on $B_0(K)$ by

$$(\forall a \in B_0(K)) \quad I(a) = J(U^{-1}(\{a\})) \tag{4}$$

Note that I is well-defined. Clearly, it holds that $(\forall f \in L_0)\ I(U(f)) = J(f)$.

This paragraph proves that it holds that

$$(\forall a, b \in B_0(K))(\forall \alpha \in [0,1]) \quad I(\alpha a + (1-\alpha)b) = \alpha I(a) + (1-\alpha)I(b)\,. \tag{5}$$

To this end, let $f, g \in L_0$ be such that $U(f) = a$ and $U(g) = b$. Then, by the previous paragraph,

$$\begin{aligned} I(\alpha a + (1-\alpha)b) &= J(U^{-1}(\{\alpha a + (1-\alpha)b\})) \\ &= J(U^{-1}(\{\alpha U(f) + (1-\alpha)U(g)\})) \\ &= J(\alpha f + (1-\alpha)g) \\ &= \alpha J(f) + (1-\alpha)J(g) \\ &= \alpha I(a) + (1-\alpha)I(b) \end{aligned}$$

where the fourth equality holds because J is an affine function.

This paragraph proves that the functional I satisfies all the conditions of the corollary of the Riesz representation theorem (Corollary 1 in Mathematical

Appendix). (i) $I(\chi_S) = 1$. It suffices to take a such that $(\forall s)\ a(s) = u(y^*) = 1$ in Equation (4). (ii) Additivity. In Equation (5), letting $b : s \mapsto u(y_*)$ shows

$$(\forall a \in B_0(K))(\forall \alpha \in [0,1]) \quad I(\alpha a) = \alpha I(a) . \tag{6}$$

Additivity follows from this and setting $\alpha = 1/2$ in Equation (5). (iii) Monotonicity. Let $a, b \in B_0(K)$ and let $f, g \in L_0$ be such that $U(f) = a$ and $U(g) = b$. Then, it follows that

$$a \geq b \Rightarrow U(f) \geq U(g) \Rightarrow (\forall s)\ u(f(s)) \geq u(g(s))$$
$$\Rightarrow (\forall s)\ f(s) \succeq g(s) \Rightarrow f \succeq g \Rightarrow J(f) \geq J(g) \Rightarrow I(a) \geq I(b) .$$

By the previous paragraph, we may apply the corollary of the Riesz representation theorem (Corollary 1 in Mathematical Appendix). Hence, if we let $p(A) = I(\chi_A)$, it holds that

$$(\forall a) \quad I(a) = \int_S a(s)\, dp(s) ,$$

from which it follows that

$$f \succ g \Leftrightarrow J(f) > J(g) \Leftrightarrow I(U(f)) > I(U(g))$$

$$\Leftrightarrow \int_S U(f)(s)\, dp(s) > \int_S U(g)(s)\, dp(s)$$

$$\Leftrightarrow \int_S u(f(s))\, dp(s) > \int_S u(g(s))\, dp(s) .$$

■

8 Ellsberg's Paradox

An essential critique to the subjective expected utility by Savage is given by *Ellsberg's Paradox*.

Example 1 (Ellsberg, 1961). There are 90 colored balls in an urn. While 30 balls out of them are known to be red (R), the remaining 60 balls are either black (B) or white (W) and their ratio is not known. Now consider betting such that a ball is drawn from this urn and a reward is given depending on the color of that ball. In particular, suppose that there are four betting described by the table below. For example, f_1 means betting such that a reward of \$1,000 is given if a red ball is drawn but nothing is given otherwise.

	R	B	W
f_1	\$1,000	\$0	\$0
f_2	\$0	\$1,000	\$0
f_3	\$1,000	\$0	\$1,000
f_4	\$0	\$1,000	\$1,000

Consider a preference order with respect to these betting (acts) such that $f_1 \succ f_2$ and $f_4 \succ f_3$. In fact, this seems to be a plausible preference order. However, for any utility index u such that $u(\$1,000) > u(\$0)$ and any (additive) probability measure p (*i.e.*, a probability of R, B and W in a usual sense with $p(W) = 1/3$), this preference order cannot be represented by the expected utility Model. Furthermore, the situation cannot be improved even if we consider the probabilistic sophistication. (Recall that the representation by the probabilistic sophistication satisfies the monotonicity.)

A thought experiment shows that the pattern of preferences exhibited in Ellsberg's paradox is reasonable. Actually, many real experiments observe the preference pattern in the paradox. These facts suggest that the representation of human behaviors by means of *additive* probability measures has a limitation, which motivates the use of *non-additive* probability measures.

9 Non-additive Measure and Choquet Integral

9.1 Non-additive Measure

Consider a measurable space (S, Σ) consisting of a set S and an algebra defined on it, Σ. A set S may be regarded as a state space and an algebra Σ may be regarded as a set of events. A set function $v : \Sigma \to \mathbb{R}$ is called a *non-additive finite measure* or *finite capacity*[7] if it satisfies the following two conditions:

$$v(\phi) = 0 \tag{7}$$

$$(\forall A, B \in \Sigma) \quad A \subseteq B \Rightarrow v(A) \leq v(B) \tag{8}$$

From these two conditions, it follows that $(\forall A \in \Sigma)\ 0 \leq v(A) \leq v(S) < +\infty$. In what follows, we always consider a non-additive finite measure which is normalized so that

$$v(S) = 1 \tag{9}$$

holds. A set function v which satisfies Equations (7), (8) and (9) is called *non-additive probability measure* or *probability capacity*. In many cases, these are abbreviated to "non-additive measure" or "capacity."

A non-additive measure is said to be *convex* if it holds that

$$(\forall A, B \in \Sigma) \quad v(A \cup B) + v(A \cap B) \geq v(A) + v(B) . \tag{10}$$

If the converse inequality holds in (10), v is said to be *concave*. When the inequalities always holds with equalities, v is a probability measure. Also, v is said to be *convex-ranged* if the next condition holds:

$$(\forall A \in \Sigma)(\forall r \in [0, v(A)])(\exists B \in \Sigma) \quad B \subseteq A \text{ and } v(B) = r .$$

[7] A non-additive finite measure is sometimes called a *game*.

Let $P(S, \Sigma)$ be the set of all probability measures on a measurable space (S, Σ). We denote by $\text{core}(v)$ the *core* of a non-additive measure v and define it by

$$\text{core}(v) = \{ p \in P(S, \Sigma) \mid (\forall A \in \Sigma) \;\; p(A) \geq v(A) \} .$$

If a non-additive measure v is a probability measure, $\text{core}(v)$ consists only of v itself. When v is convex, $\text{core}(v)$ turns out to be nonempty.

9.2 Choquet Integral

We denote by $B(S, \Sigma)$, or more simply B, the set of all Σ-measurable and bounded real-valued functions defined on a measurable space (S, Σ). Here, a function $a : S \to \mathbb{R}$ is *Σ-measurable* if for any Borel set E on $\mathbb{R}$, $a^{-1}(E) \in \Sigma$. We denote by $B_0(S, \Sigma)$ or B_0 the subset of $B(S, \Sigma)$ consisting of functions, called *simple functions*, whose ranges are finite sets. Given a non-additive measure v, we define a (nonlinear) functional $I : B \to \mathbb{R}$ by: $(\forall a \in B)$

$$\begin{aligned} I(a) &= \int a \, dv = \int_S a(s) \, dv(s) \qquad (11) \\ &= \int_{-\infty}^{0} (v(a \geq y) - v(S)) \, dy + \int_{0}^{+\infty} v(a \geq y) \, dy \\ &= \int_{-\infty}^{0} (v(\{ s \mid a(s) \geq y \}) - v(S)) \, dy + \int_{0}^{+\infty} v(\{ s \mid a(s) \geq y \}) \, dy . \end{aligned}$$

Here, two integrals in the third line are Riemann integrals in a wide sense, which are well-defined as a finite value since $a \in B$. The functional I defined by (11) is called a *Choquet integral*.

We list some properties of Choquet integrals which follow immediately from the definitions. For a function a, if we let $\underline{a} := \inf_s a(s)$, then $a - \underline{a} \geq 0$ and it holds that

$$\int (a - \underline{a}) \, dv = \int_{0}^{+\infty} v(a - \underline{a} \geq y) \, dy = \int a \, dv - \underline{a} . \qquad (12)$$

Given $a \in B_0$, we denote it by $a = \sum_{i=1}^{k} \alpha_i \chi_{E_i}$. Here, we let $\alpha_1 \geq \alpha_2 \geq \ldots \geq \alpha_k$ and $\langle E_i \rangle_{i=1}^{k}$ is a *Σ-measurable partition* of S such that $(\forall i)$ $E_i = a^{-1}(\{\alpha_i\})$. Then, the definition of a Choquet integral and (12) imply

$$\int a \, dv = \sum_{i=1}^{k} (\alpha_i - \alpha_{i+1}) v \left(\bigcup_{j=1}^{i} E_j \right) ,$$

where $\alpha_{k+1} := 0$. The next result is essential in the following analysis:

Fact 1. *If a non-additive measure v is convex, it holds that*

$$(\forall a \in B) \quad \int a \, dv = \min \left\{ \int a \, dp \;\middle|\; p \in \text{core}(v) \right\} . \qquad (13)$$

10 Choquet Expected Utility Theory à la Schmeidler

10.1 Representation Theorem by Choquet Expected Utility

We consider a setting of Anscombe-Aumann type. Given two lottery acts f and g, f and g are said to be *co-monotonic* if there does not exist any pair of states (s,t) such that $f(s) \succ f(t)$ and $g(t) \succ g(s)$. Here, $\succ$ is the binary relation on Y induced from the binary relation on L_0. We consider the following axioms about a binary relation $\succ$ on L_0.

S1 (Ordering) $\succ$ is a preference order on L_0

S2 (Co-monotonic Additivity) For any triplet (f,g,h) any two of which are co-monotonic, it follows that: $f \succ g \Rightarrow (\forall \lambda \in (0,1))\ \lambda f + (1-\lambda)h \succ \lambda g + (1-\lambda)h$

S3 (Continuity) $f \succ g$ and $g \succ h$ $\Rightarrow (\exists \alpha, \beta \in (0,1))\ \alpha f + (1-\alpha)h \succ g$ and $g \succ \beta f + (1-\beta)h$

S4 (Monotonicity) $(\forall f, g \in L_0)\ [(\forall s \in S)\ f(s) \succeq g(s)] \Rightarrow f \succeq g$

S5 (Non-degeneracy) $(\exists f, g \in L_0)\ f \succ g$

Schmeidler (1989) proved the representation theorem by *Choquet expected utility* (CEU) under these axioms.

Theorem 7 (Representation by CEU). *A binary relation $\succ$ defined on L_0 satisfies S1, S2, S3, S4 and S5 if and only if there exist a unique non-additive measure v on (S, Σ) and an affine function u on Y which is unique up to a positive affine transformation such that*

$$f \succ g \Leftrightarrow \int_S u(f(s))\, dv(s) > \int_S u(g(s))\, dv(s)\,.$$

Proof The existence of a function on L_0 which represents $\succ$ and whose restriction on L_c is an affine function. Note that a set L_c is a mixture space and that any two elements of this set are clearly co-monotonic. Since a binary relation $\succ$ restricted on L_c satisfies Ordering, Independence and Continuity by the assumption, Theorem 2 implies that there exists an affine function on L_c which represents $\succ$. We call this function u. We now define a function $J : L_0 \to \mathbb{R}$ as follows: First, if there exists $y \in Y$ such that $f = y$, define $J(f) := u(y)$. Second, for any $f \in L_0$, we know that there exist $y^*, y_* \in Y$ such that $y^* \succeq f \succeq y_*$ by S4 and that there exists α such that $f \sim \alpha y_* + (1-\alpha)y^*$ from S2 and S3 (see Lemma 1(b)). Then, define J by $J(f) := J(\alpha y_* + (1-\alpha)y^*) = u(\alpha y_* + (1-\alpha)y^*)$. The function J thus defined turns out to be a desired function. (However, it is not guaranteed that J is an affine function on L_0.)

[Construction of a function I] Let $B_0(K)$ be the set of Σ-measurable real-valued functions defined on S whose range is a finite subset of $K := u(Y)$ and define a function $U : L_0 \to B_0(K)$ by

$$(\forall f)(\forall s) \quad U(f)(s) = u(f(s)).$$

Then, U is an onto mapping and it follows from the monotonicity that $U(f) = U(g) \Rightarrow f \sim g$. Furthermore, it turns out that, if $a, b \in B_0(K)$ are co-monotonic, $f, g \in L_0$ such that $U(f) = a$ and $U(g) = b$ are also co-monotonic, and that $(\forall \alpha \in [0,1])\ U(\alpha f + (1-\alpha)g) = \alpha U(f) + (1-\alpha)U(g)$ by the affinity of u. Then, define the functional I on $B_0(K)$ by

$$(\forall a \in B_0(K)) \quad I(a) = J(U^{-1}(\{a\})).$$

By the remark just mentioned, I is well-defined.

The proof that I satisfies all the conditions of the representation theorem of Choquet integrals (Theorem 16 in Mathematical Appendix)] (i) Let $\lambda \in K$ and let y be such that $u(y) = \lambda$. Then, that $I(\lambda \chi_S) = J(y) = u(y) = \lambda$ follows because $U(y) = u(y)\chi_S = \lambda \chi_S$. (ii) Let $a, b, c \in B_0(K)$ be such that any two of them are co-monotonic and let $f, g, h \in L_0$ be such that $U(f) = a$, $U(g) = b$ and $U(h) = c$. Then, by the previous paragraph, it holds that for any $\alpha \in (0,1)$,

$$\begin{aligned} I(a) > I(b) &\Rightarrow J(f) > J(g) \\ &\Rightarrow f \succ g \\ &\Rightarrow \alpha f + (1-\alpha)h \succ \alpha g + (1-\alpha)h \\ &\Rightarrow J(\alpha f + (1-\alpha)h) > J(\alpha g + (1-\alpha)h) \\ &\Rightarrow I(\alpha U(f) + (1-\alpha)U(h)) > I(\alpha U(g) + (1-\alpha)U(h)) \\ &\Rightarrow I(\alpha a + (1-\alpha)c) > I(\alpha b + (1-\alpha)c). \end{aligned}$$

(iii) Given functions $a, b \in B_0(K)$, let $f, g \in L_0$ be such that $U(f) = a$ and $U(g) = b$. Then, it holds that

$$\begin{aligned} &a \geq b \ \Rightarrow \ U(f) \geq U(g) \ \Rightarrow \ (\forall s)\ u(f(s)) \geq u(g(s)) \\ &\Rightarrow (\forall s)\ f(s) \succeq g(s) \ \Rightarrow \ f \succeq g \ \Rightarrow \ J(f) \geq J(g) \ \Rightarrow \ I(a) \geq I(b). \end{aligned}$$

Proof of the sufficiency of the axioms If we let $v(A) = I(\chi_A)$, the representation theorem of Choquet integrals (Theorem 16 in Mathematical Appendix) implies that

$$(\forall a) \quad I(a) = \int_S a(s)\, dv(s).$$

From this, it follows that

$$f \succ g \ \Leftrightarrow \ J(f) > J(g) \ \Leftrightarrow \ I(U(f)) > I(U(g))$$

$$\Leftrightarrow \int_S U(f)(s)\, dv(s) > \int_S U(g)(s)\, dv(s)$$

$$\Leftrightarrow \int_S u(f(s))\,dv(s) > \int_S u(g(s))\,dv(s)\,.$$

■

10.2 Rank-Dependent Expected Utility

A binary relation $\succ$ on F_0 is said to be represented by a *rank-dependent expected utility* if there exist a convex-ranged probability measure p on (S, Σ), a real-valued function u on X which is unique up to a positive affine transformation and a strictly increasing function $\phi : [0,1] \to [0,1]$ such that

$$f \succ g \Leftrightarrow \int_S u(f(s))\,d(\phi \circ p)(s) > \int_S u(g(s))\,d(\phi \circ p)(s)\,,$$

where the integrals are Choquet integrals. The next theorem follows immediately from the representation theorem by Scott. (For the weak additivity, see Mathematical Appendix.)

Theorem 8. *Suppose that a binary relation $\succ$ on L_0 satisfies Axioms S1-S5. Then, if the non-additive measure v whose existence is guaranteed by the previous theorem is weakly additive, $\succ$ is represented by the rank-dependent expected utility.*

10.3 Uncertainty Aversion

A binary relation $\succ$ on L_0 is said to be *uncertainty averse* if it satisfies

$$(\forall f, g \in L_0)(\forall \alpha \in [0,1]) \quad f \succeq g \Rightarrow \alpha f + (1-\alpha) g \succeq g\,.$$

When $\succ$ is a preference relation, this condition is equivalent to

$$(\forall f, g, h \in L_0)(\forall \alpha \in [0,1]) \quad [f \succeq h \text{ and } g \succeq h] \Rightarrow \alpha f + (1-\alpha) g \succeq h\,.$$

Uncertainty aversion takes the form of quasi-concavity and intuitively means that smoothing utility distributions makes the decision maker better off.

Theorem 9. *Suppose that a binary relation $\succ$ satisfies Axioms S1, S2, S3, S4 and S5, and let v be the non-additive measure derived from Theorem 7. Then, v is convex if and only if $\succ$ is uncertainty averse.*

Proof This paragraph proves that the uncertainty aversion and the next condition are equivalent:

$$(\forall a, b \in B_0(K))(\forall \alpha \in [0,1]) \quad I(\alpha a + (1-\alpha) b) \geq \min\{I(a), I(b)\}\,. \tag{14}$$

Assume the uncertainty aversion and that $I(a) \geq I(b)$ without loss of generality. Let f and g be such that $U(f) = a$ and $U(g) = b$. Then, $f \succeq g$ since $J(f) = I(a) \geq I(b) = J(g)$, and this and the uncertainty aversion imply that for any α, it holds that $\alpha f + (1-\alpha)g \succeq g$. This in turn implies that $I(\alpha a + (1-\alpha)b) = I(\alpha U(f) + (1-\alpha)U(g)) = J(\alpha f + (1-\alpha)g) \geq J(g) = I(b)$ and then (14) follows. Conversely, assume that (14) holds and assume that $f \succeq g$. Then, since $I(U(f)) = J(f) \geq J(g) = I(U(g))$, it holds that $J(\alpha f + (1-\alpha)g) = I(\alpha U(f) + (1-\alpha)U(g)) \geq I(U(g)) = J(g)$, and hence, the uncertainty aversion follows.

(Necessity of the convexity) Assume the uncertainty aversion. By the previous paragraph, it follows that

$$(\forall a, b \in B_0(K))(\forall \alpha \in [0,1]) \quad I(a) = I(b) \Rightarrow I(\alpha a + (1-\alpha)b) \geq I(a). \quad (15)$$

This and the positive homogeneity of I imply that

$$(\forall a, b \in B_0(K)) \quad I(a) = I(b) \Rightarrow I(a+b) \geq I(a) + I(b). \quad (16)$$

In the following, we show that this implies the convexity of v. Assume that $v(E) \geq v(F)$ without loss of generality, and let $\gamma \geq 1$ be such that $v(E) = \gamma v(F)$. Then, since $I(\chi_E) = v(E) = \gamma v(F) = I(\gamma \chi_F)$, (16) implies that

$$I(\chi_E + \gamma \chi_F) \geq v(E) + \gamma v(F). \quad (17)$$

On the other hand, since $\chi_E + \gamma\chi_F = \chi_{E\cap F} + (\gamma-1)\chi_F + \chi_{E\cup F}$, the co-monotonic additivity of I implies that $I(\chi_E + \gamma\chi_F) = v(E\cap F) + (\gamma-1)v(F) + v(E\cup F)$. The convexity of v follows from this and (17).

(Sufficiency of the convexity) If a non-additive measure v is convex, Fact 10 in Mathematical Appendix implies that

$$(\forall a, b \in B_0(K)) \quad I(a+b) \geq I(a) + I(b). \quad (18)$$

By this and the positive homogeneity of I, it holds that

$$(\forall a, b \in B_0(K))(\forall \alpha \in [0,1]) \quad I(\alpha a + (1-\alpha)b) \geq \alpha I(a) + (1-\alpha)I(b). \quad (19)$$

Then, it immediately follows that (14) holds. ■

10.4 Ellsberg's Paradox: Reconsidered

Recall Ellsberg's paradox. Define a set function v as follows:

$$v(\phi) = 0, \quad v(\{R\}) = \tfrac{1}{3}, \quad v(\{B\}) = v(\{W\}) = \tfrac{1}{9},$$
$$v(\{R,B\}) = v(\{R,W\}) = \tfrac{4}{9}, \quad v(\{B,W\}) = \tfrac{2}{3} \quad \text{and} \quad v(\{R,B,W\}) = 1.$$

The set function thus defined is clearly non-additive. For example,

$$v(\{B,W\}) \neq v(\{B\}) + v(\{W\})$$

In fact, it can be easily verified that v is a *convex* non-additive measure. Then, for any u such that $u(\$1,000) > u(\$0)$ (*i.e.*, if the act which certainly guarantees $\$1,000$ is strictly preferred to the one which always gives nothing), a preference in Ellsberg's paradox can be represented by CEU with v defined above.

11 Maximin Expected Utility Theory à la Gilboa-Schmeidler

11.1 Representation Theorem of Multiple Priors

Consider the following axioms with respect to a binary relation $\succ$ on L_0 under a setting of Anscombe-Aumann type:

GS1 (Ordering) $\succ$ is a preference order on L_0

GS2 (C-Independence) $(\forall f, g \in L_0)(\forall h \in L_c)(\forall \lambda \in (0,1))$
$f \succ g \;\Leftrightarrow\; \lambda f + (1-\lambda)h \succ \lambda g + (1-\lambda)h$

GS3 (Continuity) $f \succ g$ and $g \succ h$
$\Rightarrow\; (\exists \alpha, \beta \in (0,1))\;\; \alpha f + (1-\alpha)h \succ g$ and
$g \succ \beta f + (1-\beta)h$

GS4 (Monotonicity) $(\forall f, g \in L_0)\;\; [(\forall s \in S)\; f(s) \succeq g(s)]$
$\Rightarrow\; f \succeq g$

GS5 (Non-degeneracy) $(\exists f, g \in L_0)\;\; f \succ g$

GS6 (Uncertainty Aversion) $(\forall f, g \in L_0)(\forall \alpha \in [0,1])$
$f \sim g \;\Rightarrow\; \alpha f + (1-\alpha)g \succeq f$

Gilboa and Schmeidler (1989) proved a representation theorem by *maximin expected utility* (MMEU) under these axioms. In the theorem, the set C is closed with respect to the weak* topology on the set of probability measures defined on (S, Σ). And hence, the minimum always exists.

Theorem 10 (Representation by MMEU). *A binary relation $\succ$ on L_0 satisfies GS1, GS2, GS3, GS4, GS5 and GS6 if and only if there exist a unique nonempty convex subset C of probability measures on (S, Σ) and an affine function u on Y which is unique up to a positive affine transformation such that*

$$f \succ g \;\Leftrightarrow$$
$$\min\left\{ \int_S u(f(s))\, dp(s) \,\middle|\, p \in C \right\} > \min\left\{ \int_S u(f(s))\, dp(s) \,\middle|\, p \in C \right\}.$$

Proof (Sketch) Similarly to the proof of Schmeidler's theorem (Theorem 7), it can be shown that there exists a function which represents $\succ$ on L_0 whose restriction on L_c is an affine function. Furthermore, if we define a function $U : L_0 \to B_0(K)$ by $(\forall f)(\forall s)\; U(f)(s) := u(f(s))$ and if we define a functional I on a functional space $B_0(K)$ by $(\forall a \in B_0(K))\; I(a) := J(U^{-1}(\{a\}))$, it can be shown similarly that I is well-defined and that $I(\chi_S) = 1$ by an appropriate normalization of u.

It can be proved that the functional I thus defined satisfies for any $a, b \in B_0$, for any $\alpha \geq 0$ and for any $\gamma \in \mathbb{R}$,

Monotonicity	$a \geq b \Rightarrow I(a) \geq I(b)$
Positive Homogeneity	$I(\alpha a) = \alpha I(a)$
Super Additivity	$I(a+b) \geq I(a) + I(b)$
C-Additivity	$I(a + \gamma\chi_S) = I(a) + I(\gamma\chi_S)$

and that I can be extended to B so that it should satisfy these conditions.

Any functional I which satisfies Monotonicity, Positive Homogeneity, Super Additivity, C-independence and $I(\chi_S) = 1$ can be shown to be represented by

$$(\forall a \in B) \quad I(a) = \min\left\{ \int_S a(s)\, dp(s) \,\middle|\, p \in C \right\}$$

for some closed convex set C of probability measures, from which the sufficiency of the axioms follows. ■

11.2 Ellsberg's Paradox: Re-reconsidered

Recall Ellsberg's paradox again. Consider the probabilities with respect to Red (R), Black (B) and White (W), $\boldsymbol{p} = \left(\frac{1}{3}, \frac{1}{9}, \frac{5}{9}\right)$ and $\boldsymbol{q} = \left(\frac{1}{3}, \frac{5}{9}, \frac{1}{9}\right)$, and denote corresponding (additive) probability measures by p and q. Then, for any u such that $u(\$1,000) > u(\$0)$ (*i.e.*, if the act which certainly guarantees \$1,000 is strictly preferred to the one which always gives nothing), a preference order in Ellsberg's paradox can be represented by MMEU with C defined as the convex hull of $\{p, q\}$.

12 Some Remarks

12.1 A Relation between CEU and MMEU

If a non-additive measure v is convex, it follows from Fact 1 that

$$\int u(f(s))\, dv(s) = \min\left\{ \int u(f(s)) dp(s) \,\middle|\, p \in \text{core}(v) \right\}.$$

Therefore, <u>if v is convex</u> (*i.e.*, if the preference order is uncertainty-averse), CEU is a special case of MMEU with $C = \text{core}(v)$.

However, the converse is not necessarily correct. Given an arbitrary closed convex set C of probability measures, if there exists a non-additive measure v such that

$$(\forall a \in B) \quad \int a(s)\, dv(s) = \min\left\{ \int a(s)\, dp(s) \,\middle|\, p \in C \right\},$$

it must hold that

$$(\forall A \in \Sigma) \quad v(A) = \min\{\, p(A) \,|\, p \in C \,\}. \tag{20}$$

(This is clear if we take χ_A as a.) But, as the following two examples by Huber and Strassen (1973) show, it is not guaranteed that v generated from C by this equation is convex nor that even if so, it holds that $\mathrm{core}(v) = C$. That is, for an arbitrarily given C, it is not guaranteed that there exists a convex non-additive measure which generates C as its core.[8][9]

Example 2. Let a state space S be $S = \{1, 2, 3, 4\}$ and let $\Sigma = 2^S$. Let the (additive) probability measures p and q on a measurable space (S, Σ) be corresponding to the probabilities on S, $\boldsymbol{p} = \left(\frac{5}{10}, \frac{2}{10}, \frac{2}{10}, \frac{1}{10}\right)$ and $\boldsymbol{q} = \left(\frac{6}{10}, \frac{1}{10}, \frac{1}{10}, \frac{2}{10}\right)$, and let C be the convex hull of a set $\{p, q\}$. Define sets A and B by $A = \{1, 2\}$ and $B = \{1, 3\}$. If we calculate v from Equation (20), we obtain v such that

$$v(A) = v(B) = \tfrac{7}{10}\,, \quad v(A \cap B) = \tfrac{5}{10}\,, \quad v(A \cup B) = \tfrac{8}{10}\,.$$

However, v is not convex since $v(A \cup B) + v(A \cap B) = \frac{13}{10} < \frac{14}{10} = v(A) + v(B)$.

Example 3. Let a state space S be $S = \{1, 2, 3\}$ and let $\Sigma = 2^S$. Let the (additive) probability measures p and q on a measurable space (S, Σ) be corresponding to the probabilities on S, $\boldsymbol{p} = \left(\frac{1}{2}, \frac{1}{2}, 0\right)$ and $\boldsymbol{q} = \left(\frac{2}{3}, \frac{1}{6}, \frac{1}{6}\right)$, and let C be the convex hull of a set $\{p, q\}$. If we calculate v from Equation (20), we obtain v such that

$$\begin{gathered} v(\phi) = 0\,, \quad v(\{1\}) = \tfrac{1}{2}\,, \quad v(\{2\}) = \tfrac{1}{6}\,, \quad v(\{3\}) = 0\,, \\ v(\{1,2\}) = \tfrac{5}{6}\,, \quad v(\{1,3\}) = \tfrac{1}{2}\,, \quad v(\{2,3\}) = \tfrac{1}{3}\,, \quad \text{and} \quad v(S) = 1\,. \end{gathered}$$

It can be easily verified that v is certainly a convex non-additive measure. Also, it can be seen that the core of v is given by the set of probability measures corresponding to the set of probabilities given by

$$\left\{ \left(\frac{3+t}{6}, \frac{3-t-s}{6}, \frac{s}{6} \right) \,\middle|\, s, t \in [0,1] \right\}.$$

From this, it immediately follows that C is a *proper* subset of $\mathrm{core}(v)$. (For example, consider a probability measure corresponding to a probability $\left(\frac{1}{2}, \frac{1}{3}, \frac{1}{6}\right)$.) And hence, $\mathrm{core}(v) \neq C$.

---

[8] To see this, let C be an arbitrary convex set and let v be a convex non-additive measure such that $\mathrm{core}(v) = C$. Such C and v must satisfy (20). However, v' which is generated by the right-hand side of (20) need not be equal to v since $\mathrm{core}(v')$ need not be equal to $C = \mathrm{core}(v)$ as the example shows. This is a contradiction and verifies the claim.

[9] In general, a non-additive measure v is said to be *exact* if $\mathrm{core}(v)$ is nonempty and it holds that

$$(\forall A \in \Sigma) \quad v(A) = \min\{\, p(A) \,|\, p \in \mathrm{core}(v) \,\}\,.$$

Clearly, the convexity implies the exactness.

12.2 Nonlinear Expected Utility Theory of Savage Type

We briefly mention representation theorems which use acts of Savage type. The CEU theorem with Savage acts was first proved by Gilboa (1987) and then a simplification of axioms was made by Sarin and Wakker (1992).

In Savage's theorem, a state space S must be an uncountable set. This follows from Axiom P6 and it is also clear since the probability measure derived in the process of the proof is convex-ranged. A representation theorem by SEU when a state space S is a finite set was proved by Gul (1992) and the one by CEU in this case was proved by Nakamura (1990).

On the other hand, a representation theorem by MMEU with Savage acts was proved by Casadesus-Masanell, Klibanoff and Ozdenoren (2000). This theorem holds whether a state space is a finite set or an infinite set.

12.3 Epstein's Critique

Epstein (1999) showed that assuming a preference relation represented by CEU with a convex non-additive measure is not necessary nor sufficient for explaining Ellsberg's paradox. First, define a non-additive measure v by

$$\begin{gathered} v(\phi)=0\,,\;\; v(\{R\})=\tfrac{8}{24}\,,\;\; v(\{B\})=v(\{W\})=\tfrac{7}{24}\,, \\ v(\{R,B\})=v(\{R,W\})=\tfrac{1}{2}\,,\;\; v(\{B,W\})=\tfrac{13}{24}\,,\;\; \text{and}\;\; v(\{R,B,W\})=1\,. \end{gathered}$$

Then, CEU with v thus defined can explain the paradox, but v is not convex. That is, the convexity of a non-additive measure is not a necessary condition for explaining the paradox. Next, consider a non-additive measure v defined by

$$\begin{gathered} v(\phi)=0\,,\;\; v(\{R\})=\tfrac{1}{12}\,,\;\; v(\{B\})=v(\{W\})=\tfrac{1}{6}\,, \\ v(\{R,B\})=v(\{R,W\})=\tfrac{1}{2}\,,\;\; v(\{B,W\})=\tfrac{1}{3}\,,\;\; \text{and}\;\; v(\{R,B,W\})=1\,. \end{gathered}$$

The non-additive measure v thus defined *is* convex, but CEU with this v cannot explain the paradox. Epstein concludes that if Ellsberg's paradox embodies "uncertainty" itself, it is hard to understand that CEU (and MMEU) explains behavior under uncertainty.

13 An Economic Application

13.1 Non-differentiability of Choquet Integrals

We now consider (non-)differentiability of Choquet integrals in the following sense. Suppose that a function $f : S \times \mathbb{R} \to \mathbb{R}$ satisfies the following two conditions:

$$\begin{aligned} &(\forall z \in \mathbb{R})\ f(\cdot, z) \text{ is measurable and} \\ &(\forall s \in S)\ f(s, \cdot) \text{ is a differentiable concave function.} \end{aligned}$$

Since the function $z \mapsto \int f(s,z)v(ds)$ is a concave function when v is convex by Fact 10, there exist the left and right derivatives with respect to z. However, it is not guaranteed that both coincide unless v is additive as the next example shows.

Example 4. Let $S := \{1,2\}$ and let $\Sigma := 2^S$. Define a non-additive measure v on a measurable space (S,Σ) by:

$$v(\phi) = 0\,,\quad v(\{1\}) = v(\{2\}) = \frac{1}{4}\,,\quad \text{and} \quad v(S) = 1\,.$$

Then, v is convex. Also, define a function $x : S \times \mathbb{R}_+ \to \mathbb{R}$ by:

$$x(s,z) = z^{\frac{s}{2}}\,.$$

Then, for each s, x is concave and differentiable with respect to z. Furthermore, it follows that

$$\begin{aligned}
\frac{d}{dz}\int x(s,z)v(ds) &= \frac{d}{dz}\int z^{\frac{s}{2}}v(ds) \\
&= \frac{d}{dz}\begin{cases} (z^{\frac{1}{2}} - z)\frac{1}{4} + z & \text{if} \quad z < 1 \\ (z - z^{\frac{1}{2}})\frac{1}{4} + z^{\frac{1}{2}} & \text{if} \quad z > 1 \end{cases} \\
&= \begin{cases} \frac{1}{8}z^{-\frac{1}{2}} + \frac{3}{4} & \text{if} \quad z < 1 \\ \frac{3}{8}z^{-\frac{1}{4}} & \text{if} \quad z > 1\,, \end{cases}
\end{aligned}$$

and hence, it holds that

$$\frac{d}{dz_-}\int x(s,z)v(ds)\bigg|_{z=1} = \frac{7}{8} > \frac{5}{8} = \frac{d}{dz_+}\int x(s,z)v(ds)\bigg|_{z=1}\,.$$

Given a convex non-additive measure v and a bounded measurable real-valued function x, let $\mathcal{P}(v,x)$ be the set of probability measures defined by

$$\mathcal{P}(v,x) := \arg\min\left\{\int x\,dP \,\middle|\, P \in \text{core}(v)\right\}\,. \tag{21}$$

By Fact 1, $\mathcal{P}(v,x)$ can be seen as the set of probability measures which are "equivalent" to v with respect to a calculation of the Choquet integral of x. Fact 1 also shows that $\mathcal{P}(v,x)$ is nonempty but it is not necessarily a singleton. With respect to the left and right derivatives, Aubin (1979, p.118, Proposition 6) shows that the next result holds:

Fact 2. *Assume that a non-additive measure v is convex and that a function $x : S \times \mathbb{R} \to \mathbb{R}$ satisfies the conditions described above. Then, the following holds:*

$$(\forall z) \quad \frac{d}{dz_-}\int x(s,z)v(ds) = \max\left\{\int v_2(s,z)P(ds) \,\middle|\, P \in \mathcal{P}(v, x(\cdot,z))\right\}$$

$$(\forall z) \quad \frac{d}{dz_+} \int x(s,z)v(ds) = \min\left\{ \int v_2(s,z)P(ds) \,\middle|\, P \in \mathcal{P}(v, x(\cdot,z)) \right\},$$

where $\mathcal{P}$ is defined by (21) and v_2 denotes the partial derivative of v with respect to z.

Example 5 (Continued). We apply Fact 2 to Example 4. In this example, since $\mathcal{P}(v, x(\cdot,1)) = \text{core}(v)$,

$$\begin{aligned} \frac{d}{dz_-} \int x(s,z)v(ds)\Big|_{z=1} &= \max\left\{ \int v_2(s,1)P(ds) \,\middle|\, P \in \mathcal{P}(v, x(\cdot,1)) \right\} \\ &= \max\left\{ \int \left(\frac{s}{2}\right) P(ds) \,\middle|\, P \in \text{core}(v) \right\} \\ &= \left(\frac{1}{2}\right)\left(\frac{1}{4}\right) + \left(\frac{2}{2}\right)\left(\frac{3}{4}\right) = \frac{7}{8} \end{aligned}$$

and

$$\begin{aligned} \frac{d}{dz_+} \int x(s,z)v(ds)\Big|_{z=1} &= \min\left\{ \int v_2(s,1)P(ds) \,\middle|\, P \in \mathcal{P}(v, x(\cdot,1)) \right\} \\ &= \min\left\{ \int \left(\frac{s}{2}\right) P(ds) \,\middle|\, P \in \text{core}(v) \right\} \\ &= \left(\frac{1}{2}\right)\left(\frac{3}{4}\right) + \left(\frac{2}{2}\right)\left(\frac{1}{4}\right) = \frac{5}{8}, \end{aligned}$$

which verify the result obtained in Example 4.

13.2 Portfolio Selection Model à la Dow-Werlang

Consider an optimal investment problem of an investor who is uncertainty averse. Let $W > 0$ be her initial wealth, let X be the stochastic present value of a dividend of an asset, let z be the volume of a purchase (or a sale) of the asset and let q be the price of the asset. Assume that the preference relation of the investor is represented by CEU with a non-additive measure v which is convex and that her utility index u satisfies $u' > 0$ and $u'' \leq 0$. That is, the investor is assumed to be averse with respect to uncertainty and averse or neutral with respect to risk. Define the *conjugate* v' of a non-additive measure v by $(\forall A \in \Sigma)$ $v'(A) := 1 - v(A^c)$. Note that v' is concave if v is convex. Dow and Werlang proved the next theorem.

Theorem 11 (Dow and Werlang, 1992). *Suppose it holds that*

$$\int X dv < q < \int X dv'.$$

Then, the investor does not change the current position: i.e., she does not purchase nor sell the asset. Furthermore, if it holds that

$$q < \int X dv \quad \left(\textit{resp.} \quad q > \int X dv' \right),$$

she purchases (resp. sells) the asset.

Proof We prove only the first half. The objective function of the investor is

$$\int u\,(W - qz + zX)\,dv\,.$$

By Fact 10 and the assumption that $u'' \leq 0$, it turns out that this objective function is concave in z. Hence, it is the best for the investor to keep the current position if the following holds:

$$\frac{d}{dz_+} \int u\,(W - qz + zX)\,dv \bigg|_{z=0} < 0 < \frac{d}{dz_-} \int u\,(W - qz + zX)\,dv \bigg|_{z=0}.$$

By applying Fact 2 to the right and left derivatives, it turns out that these strict inequalities are equivalent to the following series of strict inequalities:

$$\begin{aligned}
&\min\left\{ \int u'(W)(-q + X)dP \,\middle|\, P \in \mathcal{P}(v, u(W)) \right\} \\
&< 0 < \max\left\{ \int u'(W)(-q + X)dP \,\middle|\, P \in \mathcal{P}(v, u(W)) \right\} \\
\Leftrightarrow\ &\min\left\{ \int u'(W)(-q + X)dP \,\middle|\, P \in \mathrm{core}(v) \right\} \\
&< 0 < \max\left\{ \int u'(W)(-q + X)dP \,\middle|\, P \in \mathrm{core}(v) \right\} \\
\Leftrightarrow\ &\min\left\{ \int XdP \,\middle|\, P \in \mathrm{core}(v) \right\} < q < \max\left\{ \int XdP \,\middle|\, P \in \mathrm{core}(v) \right\} \\
\Leftrightarrow\ &\int X dv < q < \int X dv'
\end{aligned}$$

where the first equivalence holds since $\mathcal{P}(v, u(W)) = \mathrm{core}(v)$ because W is constant, the second equivalence holds since $u' > 0$ by the assumption and the third equivalence holds by Fact 10.[10] ∎

In this theorem, when v is additive, the bid-ask spread given by $\int X dv \leq \int X dv'$ disappears and the model is reduced to the classical one by Arrow (1965).

[10] It can be shown that if v is convex, it holds that

$$(\forall a \in B) \quad \int a\, dv' = \max\left\{ \int a\, dp \,\middle|\, p \in \mathrm{core}(v) \right\}.$$

This theorem is quite interesting in showing the existence of portfolio inertia where no trade takes place when the investor is uncertainty-averse. Since the inertia is widely observed in the real world, their model using a non-additive measure succeeds in explaining an actual phenomenon.

Mathematical Appendix

In this appendix, we summarize some mathematical results on non-additive measures and Choquet integrals, mainly the ones which are necessary for reading the main text.

In what follows, we fix a measurable space (S, Σ) consisting of a set S and an algebra Σ defined on it and fix a non-additive measure v defined on (S, Σ). We denote by $B(S, \Sigma)$ or more simply by B the set of all Σ-measurable and bounded real-valued functions defined on the measurable space (S, Σ). Also, we denote the subset of $B(S, \Sigma)$ consisting of all the simple functions by $B_0(S, \Sigma)$ or B_0. The Choquet integrals with respect to v will be defined on B or B_0 in the manner defined in the main text.

A.1. Representation Theorems by Scott and Gilboa

Given a probability measure p on the measurable space (S, Σ) and a nondecreasing function ϕ on $[0, 1]$ which satisfies that $\phi(0) = 0$ and $\phi(1) = 1$, if we define a set function $v = \phi \circ p$ by

$$(\forall A \in \Sigma) \quad v(A) = \phi \circ p(A) = \phi(p(A)), \tag{22}$$

then v is clearly a non-additive measure on (S, Σ). However, it is not the case that any non-additive measure can be decomposed like this as the next example by Chateauneuf (1991, Example 4, p.364) shows:

Example 6. Let $S := \{1, 2, 3, 4\}$ and let a set function $m : 2^S \to [0, 1]$ be defined by:

$$m(\{1\}) = m(\{3\}) = \tfrac{1}{5}; \quad m(\{2\}) = m(\{4\}) = m(\{2, 4\}) = \tfrac{1}{6};$$
$$m(S) = \tfrac{1}{10} \text{ and for any other } A \subseteq S, \ m(A) = 0.$$

Furthermore, if we define a set function $v : 2^S \to [0, 1]$ by:

$$(\forall A) \quad v(A) = \sum_{B \subseteq A} m(B),$$

then it can be easily verified that v is a convex non-additive measure. Now suppose that (22) holds for some probability measure p and for some nondecreasing function ϕ. Then, it holds that $v(\{1\}) = v(\{3\}) = \frac{1}{5} > \frac{1}{6} = v(\{2\}) = v(\{4\})$ and that $p(\{1\}) > p(\{2\})$ and $p(\{3\}) > p(\{4\})$ since ϕ is nondecreasing, and hence, it follows that $p(\{1, 3\}) > p(\{2, 4\})$. However, since $v(\{1, 3\}) = \frac{2}{5} < \frac{1}{2} = v(\{2, 4\})$, this is a contradiction.

We consider conditions under which such a decomposition is possible. In what follows, we assume that $\Sigma = 2^S$. A non-additive measure v is said to be *weakly additive* if it satisfies the next condition:

$$\begin{gathered}(\forall A, B, E, F \subseteq S)\\ E \subseteq A \cap B,\ F \subseteq (A \cup B)^c,\ v((A \backslash E) \cup F) > v((B \backslash E) \cup F)\\ \Rightarrow\ v(A) > v(B)\end{gathered}$$

Also, v is said to be *almost weakly additive* if it satisfies the condition that there exists a countable set $M \subseteq [0,1]$ such that

$$\begin{gathered}(\forall A, B, E, F \subseteq S)\\ E \subseteq A \cap B,\ F \subseteq (A \cup B)^c,\ v((A \backslash E) \cup F) > v((B \backslash E) \cup F)\\ \Rightarrow\ v(A) > v(B) \text{ or } v(A) = v(B) \in M\,.\end{gathered}$$

Furthermore, v is said to be *infinitely decomposable* if it satisfies the condition (which is omitted here) of Gilboa (1985, p.10). Then the next theorem holds.

Theorem 12 (Scott). *For any convex-ranged non-additive measure v, it is weakly additive if and only if there exists a unique strictly increasing function $\phi : [0,1] \to [0,1]$ and a unique convex-ranged probability measure p such that $v = \phi \circ p$.*

Proof (Sketch) The necessity of weak additivity follows immediately. To prove sufficiency, assume that a convex-ranged non-additive measure is weakly additive and define a binary relation $\succ_\ell$ on Σ by $(\forall A, B)\ A \succ_\ell B \Leftrightarrow v(A) > v(B)$. Then, it can be easily seen that $\succ_\ell$ is a qualitative probability. (QP4 follows from the weak additivity.) Furthermore, it can be proved that $\succ_\ell$ satisfies QP5 (omitted). Therefore, Savage's Subjective Probability Theorem implies that there exists a unique convex-ranged probability measure p on (S, Σ) such that

$$(\forall A, B) \quad p(A) > p(B) \ \Leftrightarrow\ A \succ_\ell B \ \Leftrightarrow\ v(A) > v(B)\,. \tag{23}$$

Since a probability measure p is convex-ranged, for any $r \in [0,1]$, there exists $A \in \Sigma$ such that $p(A) = r$. Denote this set by $p^{-1}(r)$ and define a function $\phi : [0,1] \to [0,1]$ by $\phi(r) = v(p^{-1}(r))$. Then, (23) implies that ϕ is well-defined and satisfies that $\phi(0) = 0$ and $\phi(1) = 1$. From this, (22) follows and the uniqueness of ϕ follows from that of p. Finally, it follows that the function ϕ is strictly increasing from (23) and since p is convex-ranged. ■

Theorem 13 (Gilboa, 1985). *For any convex-ranged non-additive measure v, it is almost weakly additive and infinitely decomposable if and only if there exists a unique nondecreasing function $\phi : [0,1] \to [0,1]$ and a unique convex-ranged probability measure p such that $v = \phi \circ p$.*

A.2. Some Results on Choquet Integrals

In what follows, v denotes a non-additive measure defined on (S, Σ) and I denotes a Choquet integral on $B(S, \Sigma)$ or $B_0(S, \Sigma)$ defined with respect to v. The next result follows immediately from the monotonicity of v.

Fact 3 (Monotonicity). $(\forall a, b \in B) \quad a \geq b \Rightarrow I(a) \geq I(b)$

For any pair of functions $a, b \in B$, they are said to be *co-monotonic* if it holds that

$$(\forall s, t \in S) \quad (a(s) - a(t))(b(s) - b(t)) \geq 0\,.$$

The next result also follows immediately from the definition.

Fact 4 (Co-monotonicity of simple functions). *For any pair of functions $b, c \in B_0$, they are co-monotonic if and only if the next condition holds: There exist a natural number k, a Σ-measurable partition $\langle E_i \rangle_{i=1}^k$ and two k-dimensional vectors $(\beta_1, \beta_2, \ldots, \beta_k)$ and $(\gamma_1, \gamma_2, \ldots, \gamma_k)$ such that $\beta_1 \geq \beta_2 \geq \cdots \geq \beta_k$ and $\gamma_1 \geq \gamma_2 \geq \cdots \geq \gamma_k$ and such that*

$$b = \sum_{i=1}^{k} \beta_i \chi_{E_i} \quad \textit{and} \quad c = \sum_{i=1}^{k} \gamma_i \chi_{E_i}\,.$$

Fact 5 (Continuity). *For any function $a \in B$, there exists a sequence of pairs of co-monotonic simple functions, (a_n, b_n), which satisfies that*

$$(\forall n) \quad a_n \leq a_{n+1} \leq \cdots \leq a \leq \cdots \leq b_{n+1} \leq b_n \quad \textit{and}$$
$$\lim_{n \to \infty} I(a_n) = I(a) = \lim_{n \to \infty} I(b_n)\,.$$

Fact 6 (Co-monotonic Additivity). *For any pair of functions $a, b \in B$, if a and b are co-monotonic, it holds that $I(a + b) = I(a) + I(b)$.*

Fact 7 (Positive Homogeneity). $(\forall a \in B)(\forall \lambda \geq 0) \quad I(\lambda a) = \lambda I(a)$.

A.3. Representation by Choquet Integrals

For a functional $I : B \to \mathbb{R}$ on a measurable space $B(S, \Sigma)$, it is said to satisfy the *co-monotonic additivity* if it holds that

$$(\forall a, b \in B) \quad a \text{ and } b \text{ are co-monotonic} \Rightarrow I(a + b) = I(a) + I(b)$$

and it is said to satisfy the *additivity* if it holds that

$$(\forall a, b \in B) \quad I(a + b) = I(a) + I(b)\,.$$

Also, I is said to satisfy the *monotonicity* if it holds that

$$(\forall f, g \in B) \quad a \geq b \Rightarrow I(a) \geq I(b)\,.$$

If a functional I satisfies the co-monotonic additivity, it holds that $(\forall a \in B)(\forall r \in \mathbb{Q}_+)$ $I(ra) = rI(a)$. For any $r = m/n$ $(m, n \in \mathbb{N})$, the co-monotonic additivity implies that $nI((m/n)a) = I(n(m/n)a) = I(ma) = mI(a)$. In particular, $I(0) = 0$ holds. Furthermore, the following lemmas follow.

Lemma 2. *If a functional $I : B \to \mathbb{R}$ satisfies the co-monotonic additivity and monotonicity, I is continuous with respect to the norm topology on B.*

Lemma 3. *If a functional $I : B \to \mathbb{R}$ satisfies the co-monotonic additivity and monotonicity, I satisfies the positive homogeneity:*

$$(\forall a \in B)(\forall \lambda \geq 0) \quad I(\lambda a) = \lambda I(a)$$

In addition, if I satisfies the additivity, I satisfies the homogeneity:

$$(\forall a \in B)(\forall \lambda \in \mathbb{R}) \quad I(\lambda a) = \lambda I(a)\,.$$

Proof The positive homogeneity follows from the homogeneity with respect to the rational numbers and the norm continuity. Also, if I is additive, it holds that $I(a) = -I(-a)$ by $0 = I(a - a) = I(a) + I(-a)$. The conclusion follows from this and the positive homogeneity. ∎

Theorem 14 (Riesz Representation Theorem). *For a linear functional $I : B \to \mathbb{R}$ which is norm continuous and satisfies that $I(\chi_S) = 1$, it holds that*

$$(\forall a \in B) \quad I(a) = \int_S a(s)\, dp(s)\,. \tag{24}$$

Here, p is a probability measure on (S, Σ) defined by $(\forall A \in \Sigma)\ p(A) = I(\chi_A)$.[11]

Let K be a convex set which satisfies that $[-1, 1] \subseteq K \subseteq \mathbb{R}$ and denote the subset of B (or B_0) consisting of K-valued functions by $B(K)$ (or $B_0(K)$). Then, the next corollary holds.

Corollary 1. *For a functional $I : B(K) \to \mathbb{R}$ which is additive and monotonic and satisfies that $I(\chi_S) = 1$, (24) holds.*

Proof By the additivity and monotonicity, I is continuous with respect to the norm topology (Lemma 2) and satisfies the homogeneity (Lemma 3). By the homogeneity, I can be extended to B. Since I thus extended is a continuous linear functional on B, the result follows from Riesz Representation Theorem. ∎

Theorem 15 (Schmeidler's (1986) Representation Theorem). *Suppose that $I : B \to \mathbb{R}$ is a functional which satisfies that $I(\chi_S) = 1$. Then, if I satisfies the co-monotonic additivity and monotonicity, I can be represented by the Choquet integral with respect to the non-additive measure v defined by $(\forall E \in \Sigma)\ v(E) = I(\chi_E)$.*

[11] For integrals with respect to probability measures and a proof of Riesz Representation Theorem, see, for example, Dunford and Schwartz (1988) and Rao and Rao (1983).

We already pointed out that Choquet integrals satisfy the co-monotonic additivity and monotonicity. This theorem is its "converse." The theorems where B is replaced by B_0 in the above and the next theorem where $B(K)$ is replaced by $B_0(K)$ also hold.

Theorem 16 (Schmeidler, 1986). *Suppose that a function $I : B(K) \to \mathbb{R}$ satisfy the following three conditions: (i) $(\forall \lambda \in K)\ I(\lambda \chi_S) = \lambda$; (ii) For any triplet of functions, (a, b, c), any two of which are co-monotonic, if $I(a) > I(b)$, then it holds that $(\forall \alpha \in (0,1))\ I(\alpha a + (1-\alpha)c) > I(\alpha b + (1-\alpha)c)$; and (iii) $a \geq b \Rightarrow I(a) \geq I(b)$. Then, the function I can be represented by the Choquet integral with respect to the non-additive measure v defined by $(\forall E \in \Sigma)\ v(E) = I(\chi_E)$.*

A.4. Some Results on Convexity of Non-additive Measures

Fact 8. *Given a probability measure p on a measurable space (S, Σ) and a non-decreasing convex function ϕ on $[0,1]$ which satisfies $\phi(0) = 0$ and $\phi(1) = 1$, then the set function v defined by $v = \phi \circ p$ is a convex non-additive measure.*

Fact 9 (Shapley, 1971). *If a non-additive measure v is convex, then* core(v) *is nonempty.*

Fact 10 (Schmeidler, 1986). *For a Choquet integral with respect to a non-additive measure v, the following three conditions are equivalent:*

(i) *v is convex*

(ii) $(\forall a \in B)\quad \int a\, dv = \min \left\{ \int a\, dp \,\middle|\, p \in \text{core}(v) \right\}$

(iii) $(\forall b, c \in B)\quad \int (b + c)\, dv \geq \int b\, dv + \int c\, dv$

The three conditions where the convexity is replaced by the concavity, "min" is replaced by "max" and the direction of the inequality is reversed in each of (i)-(iii) are also equivalent.

References

Anscombe, F.J., Aumann, R.J.: A Definition of Subjective Probability. Annals of Mathematical Statistics 34, 199–205 (1963)

Arrow, K.J.: The Theory of Risk Aversion. Aspects of the Theory of Risk Bearing. ch. 2, Yrjo Jahnsonin Saatio, Helsinki (1965)

Aubin, J.P.: Mathematical Methods of Game and Economic Theory. North-Holland, Amsterdam (1979)

Casadesus-Masanell, R., Klibanoff, P., Ozdenoren, E.: Maxmin Expected Utility over Savage Acts with a Set of Priors. Journal of Economic Theory 92, 35–65 (2000)

Chateauneuf, A.: On the Use of Capacities in Modeling Uncertainty Aversion and Risk Aversion. Journal of Mathematical Economics 20, 343–369 (1991)

Dow, J., Werlang, S.R.C.: Uncertainty Aversion, Risk Aversion, and the Optimal Choice of Portfolio. Econometrica 60, 197–204 (1992)

Dunford, N., Schwartz, J.T.: Linear Operator Part I: General Theory. Wiley Classics Library (1988)
Ellsberg, D.: Ambiguity, and the Savage Axioms. Quarterly Journal of Economics 75, 643–669 (1961)
Epstein, L.G.: A Definition of Uncertainty Aversion. Review of Economic Studies 66, 579–608 (1999)
Fishburn, P.C.: Utility Thoery for Decision Making. Wiley, New York (1970)
Gilboa, I.: Subjective Distortions of Probabilities and Non-Additive Probabilities, Working Paper. The Foerder Institute for Economic Research, Tel Aviv University (1985)
Gilboa, I.: Expected Utility with Purely Subjective Non-Additive Probabilities. Journal of Mathematical Economics 16, 65–88 (1987)
Gilboa, I.: Theory of Decision under Uncertainty. Cambridge University Press (2009)
Gilboa, I., Schmeidler, D.: Maxmin Expected Utility with Non-Unique Prior. Journal of Mathematical Economics 18, 141–153 (1989)
Gul, F.: Savage's Theorem with a Finite Number of States. Journal of Economic Theory 57, 99–110 (1992)
Herstein, I.N., Milnor, J.: An Axiomatic Approach to Measurable Utility. Econometrica 21, 291–297 (1953)
Huber, P.J., Strassen, V.: Minimax Tests and the Neyman-Pearson Lemma for Capacities. The Annals of Statistics 1, 251–263 (1973)
Kreps, D.: Notes on the Theory of Choice. Westview Press, Boulder (1988)
Machina, M.J., Schmeidler, D.: A More Robust Definition of Subjective Probability. Econometrica 60, 745–780 (1992)
Nakamura, Y.: Subjective Expected Utility with Non-additive Probabilities on Finite State Spaces. Journal of Economic Theory 51, 346–366 (1990)
von Neumann, J., Morgenstern, O.: Theory of Games and Economic Behavior, 2nd edn. Princeton University Press, Princeton (1947)
Rao, K.P.S.B., Rao, M.B.: Theory of Charges. Academic Press (1983)
Sarin, R., Wakker, P.: A Simple Axiomatization of Nonadditive Expected Utility. Econometrica 60, 1255–1272 (1992)
Savage, L.: The Foundations of Statistics. John Wiley, New York (1954)
Schmeidler, D.: Integral Representation without Additivity. Proceedings of the American Mathematical Society 97, 255–261 (1986)
Schmeidler, D.: Subjective Probability and Expected Utility without Additivity. Econometrica 57, 571–587 (1989)
Shapley, L.S.: Cores of Convex Games. International Journal of Game Theory 1, 11–26 (1971)

Cooperative Game as Non-Additive Measure

Katsushige Fujimoto

College of Symbiotic Systems Science, Fukushima University,
1 Kanayagawa Fukushima, 906-1296, Japan
fujimoto@sss.fukushima-u.ac.jp

Abstract. This chapter surveys cooperative game theory as an important application based on non-additive measures. In ordinary cooperative game theory, it is implicitly assumed that all coalitions of N can be formed; however, this is in general not the case. Let us elaborate on this, and distinguish several cases: 1) *Some coalitions may not be meaningful.* 2) *Coalitions may not be "black and white"*. In order to deal with such situations, various generalizations/extensions of the theory have been proposed, e.g., *bi-cooperative games, games on networks, games on combinatorial structures*. We give a survey on values and interaction indices for these extended cooperative game theory.

Keywords: cooperative game, bi-cooperative game, network, combinatorial structure, value, interaction index.

1 Introduction

Measure is one of the most important concepts in mathematics and so is the integral with respect to a measure. They have many applications in economics, engineering, and many other fields, and one of their main characteristics is additivity. This is very effective and convenient, but often too inflexible or too rigid. As a solution to the rigidness problem, several approaches based on non-additive measures have been proposed in various fields. The non-additivity can represent *interaction phenomena* among elements to be measured.

Let N be a finite set and v a set function (non-additive measure) on 2^N. Given a subset $S \subseteq N$, the precise meaning of the quantity $v(S)$ depends on the kind of intended application or domain [22]:

N **is the set of states of nature.** Then $S \subseteq N$ is an *event* in decision under uncertainty or under risk, and $v(S)$ represents the degree of certainty, belief, etc.

N **is a the set of criteria, or attributes.** Then $S \subseteq N$ is a group of criteria (or attributes) in multi-criteria (or multi-attributes) decision making, and $v(S)$ represents the degree of importance of S for making decision.

N **is the set of voters, political parties.** Then $S \subseteq N$ is called a *coalition* in voting situations, and $v(S) = 1$ iff bill passes when coalition S votes in favor of the bill, and $v(S) = 0$ else.

N **is the set of players, agents, companies, etc.** Then $S \subseteq N$ is also called a *coalition* in cooperative game theory, and $v(S)$ is the worth (or payoff,

V. Torra, Y. Narukawa, and M. Sugeno (eds.), *Non-Additive Measures*,
Studies in Fuzziness and Soft Computing 310,
DOI: 10.1007/978-3-319-03155-2_6,

or income, etc.) won by S if all members in S agree to cooperate, and the other ones do not.

In the current chapter, we discuss and focus on cooperative games as an application based on non-additive measures.

2 Ordinary Cooperative Game

2.1 Definitions and Several Representations of Cooperative Games

Definition 1 (cooperative game). The function v that assigns to every coalition $S \subseteq N$ its value or worth $v(S)$ is commonly referred to as the *characteristic function*. It is always assumed that $v(\emptyset) = 0$. A pair (N, v) consisting of a player set N and a characteristic function v constitutes a *cooperative game* or *coalitional game*. These games are also referred to as *TU games*, where TU stands for *transferable utility* (We often identify (N, v) with v). Sometimes, we want to focus on only a few of the players involved in a cooperative game (N, v). For a coalition $S \subseteq N$, $v|_S$ denotes the restriction of the characteristic function v to the player set S, i.e., $v|_S(T) = v(T)$ for each coalition $T \subseteq S$. Then, $(S, v|_S)$ is called a *subgame* of the game (N, v).

In order to avoid a heavy notation, we will often omit braces for singletons, e.g., by writing $v(i)$, $N \setminus i$ instead of $v(\{i\})$, $N \setminus \{i\}$. Similarly, for pairs, we will write ij instead of $\{i, j\}$. Furthermore, cardinalities of coalitions $S, T, \ldots,$ will often be denoted by the corresponding lower case letters $s, t, \ldots,$ otherwise by the standard notation $|S|, |T|$, etc.

The set of all cooperative game with player set N will be denoted by $\mathcal{G}^N$. The set $\mathcal{G}^N$ is a $(2^n - 1)$-dimensional linear space.

Definition 2 (unanimity game). For each $T \in 2^N$, the *unanimity game* $u_T \in \mathcal{G}^N$ is defined by

$$u_T(S) := \begin{cases} 1 & \text{if } S \supseteq T, \\ 0 & \text{otherwise.} \end{cases}$$

The set $\{u_T \mid T \in 2^N \setminus \{\emptyset\}\}$ is a basis of the linear space $\mathcal{G}^N$. For any game $v \in \mathcal{G}^N$, we have

$$v(S) = \sum_{\emptyset \neq T \subseteq N} c_T(v)\, u_T(S) \qquad \forall S \in 2^N \setminus \{\emptyset\},$$

where

$$c_T(v) = \sum_{R \subseteq T} (-1)^{|T \setminus R|} v(R) \qquad \forall T \in 2^N \setminus \{\emptyset\}.$$

Then, $\{c_T\}_{T \in 2^N \setminus \{\emptyset\}}$ is called *unanimity coefficients* or *Harsanyi dividends* [33] of v.

Definition 3 (the Möbius transforms [56]). Let $(P, \leq)$ be a poset. For a function $f : P \to \mathbb{R}$, the *Möbius transform* Δ^f of f is the unique solution of the equation:

$$f(x) = \sum_{y \leq x} \Delta^f(y) \quad \forall x \in P,$$

given by

$$\Delta^f(x) = \sum_{y \le x} \mu(y, x) f(y), \quad x \in P,$$

where μ is the so-called *Möbius function* on P and given by

$$\mu(y, x) = \begin{cases} 1 & \text{if } x = y, \\ -\sum_{y \le z < x} \mu(y, z) & \text{if } y < x, \\ 0 & \text{otherwise.} \end{cases}$$

Now, considering a pair $(2^N, \subseteq)$ as a poset and a characteristic function $v : 2^N \to \mathbb{R}$ of cooperative game (N, v), then the *Möbius transform* of the game v on the poset $(2^N, \subseteq)$, is obtained as

$$\Delta^v(T) := \sum_{R \subseteq T} (-1)^{|T \setminus R|} v(R) \qquad \forall T \in 2^N.$$

That is, the concept of Möbius transform fits with of Harsanyi dividends in cooperative game theory. i.e., $c_T(v) = \Delta^v(T)$ for any $T \in 2^N \setminus \{\emptyset\}$. Inversely,

$$v(S) = \sum_{T \subseteq S} \Delta^v(T) \qquad \forall S \in 2^N.$$

Here, the Möbius transform or Harsanyi dividends of cooperative games can be interpreted as follows:

> *The Möbius transform is vanishing at the empty set, its worth $v(i)$ for every singleton $i \in N$, while recursively, the Möbius transform of every coalition of at least two players is equal to its worth minus the sum of the Möbius transforms of all its proper subcoalitions. In this sense, the Möbius transform of a coalition S can be interpreted as an extra contribution of cooperation among the players in S that they did not already achieve by smaller coalitions.*

Definition 4 (multilinear extension). Let I^N be the n-dimensional unit hyper cube, i.e.,

$$I^N := \{(x_1, \cdots, x_n) \in \mathbb{R}^n \mid 0 \le x_i \le 1, \ \forall i \in N = \{1, \cdots, n\}\}.$$

The extreme points of I^N are the vectors χ_S, $S \subseteq N$, where

$$(\chi_S)_i = \begin{cases} 1 & \text{if } i \in S, \\ 0 & \text{otherwise.} \end{cases}$$

So, we notice that $v \in \mathcal{G}^N$ determines a real function $\bar{v}$ on the corners of I^v by

$$\bar{v}(\chi_S) = v(S) \quad \forall S \subseteq N.$$

Hence, $\bar{v}$ may be extended to I^N by

$$\bar{v}(\boldsymbol{x}) = \sum_{T \subseteq N} \left(\prod_{i \in T} x_i \prod_{i \in N \setminus S} (1 - x_i) \right) v(T) \quad \forall \boldsymbol{x} \in I^N.$$

Then, this function $\bar{v} : I^N \to \mathbb{R}$ is called the *multilinear extension* (MLE) of v. The multilinear extension of v can also be represented via the Harsanyi dividends (Möbius transform) as follows [53]:

$$\bar{v}(\boldsymbol{x}) = \sum_{T \subseteq N} \Delta^v(T) \prod_{i \in T} x_i \quad \forall \boldsymbol{x} \in I^N.$$

2.2 Intuitive Representations of Importance and Interaction

In order to intuitively approach the concept of importance of each player and of interaction among players, consider two players i and $j \in N$. Clearly, $v(i)$ is one of representations of importance of $i \in N$. An inequality

$$v(ij) > v(i) + v(j) \quad \text{(resp., <)}$$

seems to model a *positive (resp., negative) interaction* or *complementary (resp., substitutive) effect* between players i and j. However, as discussed in Grabisch and Roubens [27], the intuitive concept of interaction requires a more elaborate definition. We should not only compare $v(i)$, $v(j)$, and $v(ij)$ but also see what happens when i, j, and ij join the other coalitions. That is, we should take into account all coalitions of the form $T \cup i$, $T \cup j$, and $T \cup ij$. For a player i and a coalition $T \not\ni i$,

$$\Delta_i v(T) := v(T \cup i) - v(T) \tag{1}$$

seems to represent an index of importance of i in $T \cup i$. The equation (1) is called the *marginal contribution* of a player i to a coalition T. Then it seems natural to consider that if for T not containing i and j

$$\Delta_i v(T \cup j) > \Delta_i v(T) \quad \text{(resp., <)},$$

then i and j interact positively (resp., negatively) each other in the presence of T since the presence of j increases (resp., decreases) the marginal contribution of i to T. Then

$$\Delta_{ij} v(T) := \Delta_i v(T \cup j) - \Delta_i v(T)$$

is called the *marginal interaction* [28] between i and j in the presence of T. Note that

$$\begin{aligned}\Delta_i v(T \cup j) - \Delta_i v(T) &= v(T \cup ij) - v(T \cup i) - v(T \cup j) + v(T) \\ &= \Delta_j v(T \cup i) - \Delta_j v(T).\end{aligned}$$

For three players $i, j, k \in N$ and a coalition T not containing i, j and k, $\Delta_{\{i,j,k\}} v(T)$ can be naturally defined as

$$\Delta_{\{i,j,k\}} v(T) := \Delta_{ij} v(T \cup k) - \Delta_{ij} v(T).$$

Then we have $\Delta_{ij} v(T \cup k) - \Delta_{ij} v(T) = \Delta_{ik} v(T \cup j) - \Delta_{ik} v(T) = \Delta_{jk} v(T \cup i) - \Delta_{jk} v(T)$. Moreover, for two distinct coalitions S and $T \subseteq N \setminus S$,

$$\Delta_S v(T) := \Delta_{S \setminus i} v(T \cup i) - \Delta_{S \setminus i} v(T)$$

for $i \in S$. Then $\Delta_{S\setminus i}v(T \cup i) - \Delta_{S\setminus i}v(T) = \Delta_{S\setminus j}v(T \cup j) - \Delta_{S\setminus j}v(T)$ for any $i, j \in S$. Similarly, when, for example, $\Delta_S v(T) > 0$ (resp., <), we shall consider that players among S interact positively (resp., negatively) each other in the presence of T.

These marginal contributions and interactions can be represented through the following notion, *discrete derivative*.

Definition 5 (discrete derivative [28]). Given a game $v \in \mathcal{G}^N$ and finite coalitions $S, T \subseteq N$, we denote by $\Delta_S v(T)$ the *S-derivative* of v at T, which is recursively defined by

$$\Delta_i v(T) := v(T \cup i) - v(T \setminus i) \qquad \forall i \in N,$$

and

$$\Delta_S v(T) := \Delta_i \left(\Delta_{S\setminus i} v(T)\right) \qquad \forall i \in S,$$

with convention $\Delta_\emptyset v(T) := v(T)$.

Proposition 1 ([18,20,28]). *For any $S \subseteq N$, $T \subseteq N \setminus S$ and $v \in \mathcal{G}^N$, the S-derivative of v at T can be represented as follows:*

$$\Delta_S v(T) = \sum_{L \subseteq S} (-1)^{|S\setminus L|} v(T \cup L) = \sum_{L \subseteq T} \Delta^v(S \cup L),$$

i.e.,

$$\Delta_S v(T) = \sum_{T \subseteq L \subseteq S \cup T} (-1)^{|(S\cup T)\setminus L|} v(L) = \sum_{S \subseteq L \subseteq S \cup T} \Delta^v(L).$$

In particular, $\Delta^v(S) = \Delta_S v(\emptyset)$ for any $S \subseteq N$. Moreover, if σ is a permutation on N such that $S = \{\sigma(1), \cdots, \sigma(|S|)\}$,

$$\Delta_S v(T) = \frac{\partial^{|S|}}{\partial x_{\sigma(1)} \cdots \partial x_{\sigma(|S|)}} \bar{v}(\boldsymbol{x}) \bigg|_{\boldsymbol{x} = \chi_{S\cup T}}.$$

where $\bar{v}$ is the MLE of v.

Definition 6 (k-monotonic game [10]). Given an integer $k \geq 2$, a game $v \in \mathcal{G}^N$ is said to be *k-order monotone* (for short, *k-monotone*) if and only if, for any (at most) k coalitions $S_1, \cdots, S_k$, we have

$$v\left(\bigcup_{i=1}^{k} S_i\right) \geq \sum_{\substack{J \subseteq \{1,\cdots,k\} \\ J \neq \emptyset}} (-1)^{|J|+1} v\left(\bigcap_{i \in J} S_i\right). \tag{2}$$

It is easy to verify that k-monotonicity ($k \geq 2$) implies l-monotonicity for all integer $2 \leq l \leq k$. By extension, 1-monotonicity (which does not correspond to $k = 1$ in Eq. (2)) is defined as standard monotonicity, i.e.,

$$v(S) \leq v(T) \quad \text{whenever} \quad S \subseteq T \subseteq N.$$

A game $v \in \mathcal{G}^N$ is called *totally monotone* if Eq.(2) holds for any positive integer k. 2-Monotonic games v, i.e.,

$$v(S \cup T) \geq v(S) + v(T) - v(S \cap T) \quad \forall S, T \subseteq N,$$

are also referred to as *convex games*.

The notion of k-monotonicity can be characterized through the use of discrete derivatives as follows.

Proposition 2 ([10,18,20]). *Let v be a game on N (i.e., $v \in \mathcal{G}^N$) and k a positive integer. Then v is k order monotone if and only if*

$$\Delta_S v(T) \geq 0$$

for any $S \subseteq N$ and $T \subseteq N \setminus S$ such as $1 \leq |S| \leq k$.

Note: In evidence theory [58], belief functions $Bel : 2^N \to [0,1]$ have been introduced as totally monotonic games, i.e., whose Möbius transforms Δ^{Bel}, which are called "*basic probability assignments*", are non-negative for all events (coalitions).

2.3 The Shapley Value as an Acceptable Allocation in Cooperative Games

The players in a cooperative game are eventually interested in what they individually will get out of cooperating with the other players. How will individual players benefit from cooperation? So far, various solutions concepts and allocation rules of benefits have been proposed (see, e.g., [54]). Some of them (e.g., *the core, bargaining set, prekernel, kernel, prenucleolus, nucleolus*) are based on *domination*, and some of them (e.g., *the Shapley value*) are based on *expectation*. This subsection discusses only the Shapley value and relatives from the standpoint of Shapley's statement [59]:

> *"At the foundation of the theory of games is the assumption that the players of a game can evaluate, in their utility scale, every "prospect" that might arise a result of a play. In attempting to apply the theory to any field, one would normally expect to be permitted to include, in the class of "prospects", the prospect of having to play a game. The possibility of evaluating games is therefore of critical importance."*

A *payoff vector* or *allocation* is a vector $x = (x_1, \cdots, x_n) \in \mathbb{R}^N$ that specifies for each player $i \in N$ the profit x_i that this player can expect when he cooperates with the other players. Thus, a payoff vector $x = (x_1, \cdots, x_n)$ such as $\sum_{i \in N} x_i > v(N)$ is not feasible. That is, payoffs $x = (x_1, \cdots, x_n)$ with $\sum_{i \in N} x_i = v(N)$ are the most efficient allocations of $v(N)$. However, not all these efficient allocations will be acceptable to the players. Here, we introduce the Shapley value, which provide a priori evaluations of every cooperative game as an acceptable allocation to each player.

Definition 7 (the Shapley value [59]). The *Shapley value* $\phi : \mathcal{G}^N \to \mathbb{R}^N$ is given by

$$\phi_i(v) = \sum_{S \subseteq N \setminus i} \frac{s! \cdot (n-s-1)!}{n!} [v(S \cup i) - v(S)]$$

for any $v \in \mathcal{G}^N$ and any $i \in N$, where $\phi_i(v)$ is the i-th component of $\phi(v) \in \mathbb{R}^N$.

The Shapley value is one of the most well-known allocation rule defined as a certain type of expectation of marginal contributions for each player and characterized as the unique allocation rule satisfying the following four properties (axioms): *symmetry, efficiency, null zero,* and *additivity (low of aggregation)* [59].

Definition 8 (symmetry). Let $\Pi(N)$ denote the set of all permutations on N. If $\sigma \in \Pi(N)$, then writing $\sigma(S)$ for the image of $S \subseteq N$ under σ, i.e., $\sigma(S) := \{\sigma(i) \mid i \in S\}$, we may define the game σv by $\sigma v(\sigma(S)) = v(S)$ for all $S \subseteq N$. An allocation rule $F : \mathcal{G}^N \to \mathbb{R}^N$ is said to be *symmetry* if

$$F_{\sigma(i)}(\sigma v) = F_i(v)$$

for any $\sigma \in \Pi(N)$ and $v \in \mathcal{G}^N$, where $F_i(v)$ is the i-th component of $F(v) \in \mathbb{R}^N$.

Under symmetry allocation rules, the names of players play no role in determining the allocation to each player.

Definition 9 (efficiency). An allocation rule $F : \mathcal{G}^N \to \mathbb{R}^N$ is said to be *efficient* if

$$\sum_{i \in N} F_i(v) = v(N)$$

for any $v \in \mathcal{G}^N$, where $F_i(v)$ is the i-th component of $F(v) \in \mathbb{R}^N$.

Under efficient allocation rules, the total worth $v(N)$ is allocated to all the players.

Definition 10 (null-zero). An allocation rule $F : \mathcal{G}^N \to \mathbb{R}^N$ is said to be *null-zero* if

$$F_i(v) = 0$$

for any $v \in \mathcal{G}^N$ and $i \in N$ such that $v(S \cup i) = v(S)\ \forall S \subseteq N$, where $F_i(v)$ is the i-th component of $F(v) \in \mathbb{R}^N$.

Under null-zero allocation rules, a player who adds nothing to the worth of any coalition is allocated nothing.

Definition 11 (additivity). An allocation rule $F : \mathcal{G}^N \to \mathbb{R}^N$ is said to be *additive* if

$$F(v + w) = F(v) + F(w)$$

for any $v, w \in \mathcal{G}^N$.

Under additive allocation rules, if two allocation problems are combined into one by adding the characteristic functions, then for each player the allocation under the combined problem is the sum of the allocations under the two individual problems.

The Shapley value also can be treated as a power and/or importance index in various fields, *e.g., decision making problems [55], voting power in the council [60], etc.,* and represented via the Möbius transform and the multilinear extension of v as follows [54]:

$$\phi_i(v) = \sum_{i \in S \subseteq N} \frac{1}{s} \Delta^v(S).$$

$$\phi_i(v) = \int_0^1 \frac{\partial}{\partial x_i} \bar{v}(t, t, \cdots, t)\, dP_\phi(t),$$

where $\frac{\partial}{\partial x_i} \bar{v}(t, t, \cdots, t) := \frac{\partial}{\partial x_i} \bar{v}(x) \mid_{x=(t,t,\cdots,t)}$, $P_\phi(t) = t$ for any $t \in [0, 1]$ and the integral is to be understood in the sense of Riemann-Stieljes.

Note (the Banzhaf power index): The Banzhaf power index $\beta : \mathcal{G}^N \to \mathbb{R}^N$, defined by

$$\beta_i(v) := \sum_{S \subseteq N \setminus i} \frac{1}{2^{n-1}} [v(S \cup i) - v(S)] = \sum_{i \in T \subseteq N} \frac{1}{2^{t-1}} \Delta^v(T) \quad \forall v \in \mathcal{G}^N,$$

is also a well-known voting power index [2], which is not efficient. The Banzhaf power index also has an integral-representation as follows [13,18]:

$$\beta_i(v) = \int_0^1 \frac{\partial}{\partial x_i} \bar{v}(t, t, \cdots, t)\, dP_\beta(t) = \frac{\partial}{\partial x_i} \bar{v}(\frac{1}{2}, \frac{1}{2}, \cdots, \frac{1}{2}),$$

where $\frac{\partial}{\partial x_i} \bar{v}(t, t, \cdots, t) := \frac{\partial}{\partial x_i} \bar{v}(x) \mid_{x=(t,t,\cdots,t)}$, $P_\beta(t) = \mathbf{1}_{[0.5,1]}$ for any $t \in [0, 1]$.

2.4 Interaction Index in Cooperative Games

The study of the notion of *interaction* among players is relatively recent in the framework of cooperative game theory. The first attempt is probably due to Owen [53] for superadditive games. More developments are due to Murofushi and Soneda [50], Roubens [57], Grabisch and Roubens [27], Marichal and Roubens [49], and Fujimoto et al. [18]. The concept of interaction index, which can be seen as an extension of the notion of value, is fundamental for making it possible to measure the interaction phenomena modeled by a game on a set of players. The expression "*interaction phenomena*" refers to either *complementarity* or *redundancy* effects among players of coalitions resulting from the non additivity of the underlying game. Thus far, the notion of *interaction index* has been primarily applied to multi-criteria decision making in the framework of aggregation by the Choquet integral. In this context, it is used to appraise the overall interaction among criteria (see, e.g., [27,29,40]), thereby giving more insight into the decision problem. Other natural applications concern statistics and data analysis (see, e.g., [21,39]).

An allocation rule $\phi : \mathcal{G}^N \to \mathbb{R}^N$, e.g., the Shapley value, can be regarded as a function $F : \mathcal{G}^N \times N \to \mathbb{R}$ such that

$$F(v, i) = \phi_i(v)$$

for any $v \in \mathcal{G}^N$ and any $i \in N$. Then, setting $\mathcal{N} := 2^N \setminus \{\emptyset\}$, we define an *interaction index* as a function $I : \mathcal{G}^N \times \mathcal{N} \to \mathbb{R}$ to measure the (simultaneous) interaction among players in a cooperative game, i.e., $I(v, S)$ represents the (simultaneous) interaction among players S in playing a game v.

Definition 12 (interaction indices). The (Shapley-type) *interaction index* with respect to $S \in \mathcal{N}$ of v is defined by

$$I(v, S) := \sum_{T \subseteq N \setminus S} \frac{(n - t - s)!\, t!}{(n - s + 1)!} \Delta_S v(T).$$

This index is an extension of the Shapley value in the sense that $I(v, i)$ coincides with the Shapley value $\phi_i(v)$ of any player i.

The Shapley-type interaction index is a type of expectation of marginal interactions among the players in each coalition and characterized as the unique interaction index satisfying the following six properties (axioms): *symmetry, k-monotone positivity, dummy partnership, reduced partnership consistency, additivity, efficiency* [18].

Definition 13 (additivity). An interaction index $I : \mathcal{G}^N \times \mathcal{N} \to \mathbb{R}$ is said to be *additive* if

$$I(v + w, S) = I(v, S) + I(w, S)$$

for any $v, w \in \mathcal{G}^N$ and any $S \in \mathcal{N}$.

Definition 14 (symmetry). Let $\Pi(N)$ denote the set of all permutations on N. If $\sigma \in \Pi(N)$, then writing $\sigma(S)$ for the image of $S \subseteq N$ under σ, i.e., $\sigma(S) := \{\sigma(i) \mid i \in S\}$, we may define the game σv by $\sigma v(\sigma(S)) = v(S)$ for all $S \subseteq N$. An interaction index $I : \mathcal{G}^N \times \mathcal{N} \to \mathbb{R}$ is said to be *symmetry* if

$$I(\sigma v, \sigma(S)) = I(v, S)$$

for any $\sigma \in \Pi(N)$, any $S \in \mathcal{N}$, and any $v \in \mathcal{G}^N$.

Definition 15 (efficiency). An interaction index $I : \mathcal{G}^N \times \mathcal{N} \to \mathbb{R}$ is said to be *efficient* if

$$\sum_{i \in N} I(v, i) = v(N)$$

for any $v \in \mathcal{G}^N$.

Under efficient interaction indices, the interaction among itself is represented as its allocation for each player.

A coalition $P \in \mathcal{N}$ is said to be a *partnership* [38] in a game (N, v) if

$$v(S \cup T) = v(T)$$

for any $S \subsetneq P$ and any $T \subseteq N \setminus P$. In other words, as long as all the members of a partnership P are not all in coalition, the presence of some of them only leaves unchanged the worth of any coalition not containing elements of P. In particular $v(S) = 0$ for all $S \subsetneq P$. Thus, a partnership behaves like a single hypothetical player $[P]$, that is, the game $v \in \mathcal{G}^N$ and its reduced version $v_{[P]} \in \mathcal{G}^{(N \setminus P) \cup [P]}$, which is defined by

$$v_{[P]}(S) = \begin{cases} v(S) & \text{if } S \subseteq N \setminus P \\ v(S \cup P) & \text{if } S \ni [P] \end{cases},$$

can be considered as equivalent.

Definition 16 (reduced partnership consistency). An interaction index $I : \mathcal{G}^N \times \mathcal{N} \to \mathbb{R}$ is said to satisfy *reduced partnership consistency* property/axiom if

$$I(v, P) = I(v_{[P]}, [P])$$

for any $v \in \mathcal{G}^N$ and any partnership P in (N, v).

Recall that a partnership can be considered as behaving as a single hypothetical player. Furthermore, it is easy to verify that the marginal interaction among the players of a partnership $P \in \mathcal{N}$ in a game (N, v) in the presence of a coalition $T \subseteq N \setminus P$ is equal to the marginal contribution of P to coalition T, i.e.,

$$\Delta_P v(T) = v(T \cup P) - v(T).$$

In other words, when we measure the interaction among the players of a partnership, it is as if we were measuring the value of a hypothetical player. The *reduced partnership consistency property* then simply states that the interaction among players of a partnership P in a game (N, v) should be regarded as the value of the reduced partnership $[P]$ in the corresponding reduced game $v_{[P]}$.

A player $d \in N$ is said to be *dummy* in a game $v \in \mathcal{G}^N$ if

$$v(S \cup d) = v(S) + v(d)$$

for all $S \subseteq N \setminus d$. In other words, the marginal contribution of a dummy player $d \in N$ to any coalition $S \subseteq N \setminus d$ is simply its worth $v(d)$, i.e., there are no interaction between d and any $S \subseteq N \setminus d$. Similarly, a coalition $D \in \mathcal{N}$ is said to be *dummy* if $v(T \cup D) = v(T) + v(D)$ for any $T \subseteq N \setminus D$.

Definition 17 (dummy partnership). An interaction index $I : \mathcal{G}^N \times \mathcal{N} \to \mathbb{R}$ is said to satisfy *dummy partnership* property/axiom if the following two conditions hold:

(i) $I(v, D) = v(D)$
(ii) $I(v, S \cup D) = 0 \; \forall S (\neq \emptyset) \subseteq N \setminus D$

for any $v \in \mathcal{G}^N$ and any dummy partnership $D \in \mathcal{N}$ in (N, v).

The first part of *dummy partnership property* states that the interaction index of a dummy partnership D in a game (N, v) should be its worth since the marginal interaction among the players in D in the presence of any coalition T not containing elements of D is its worth, that is, $\Delta_D(T) = v(D)$ for any $T \subseteq N \setminus D$. The second part of the property says that there should be no simultaneous interaction among players of coalitions containing dummy partnerships since dummy partnerships behaves like a single hypothetical dummy player and he does not interact with any outsider coalition (see, [49]).

Definition 18 (k-monotone positivity). An interaction index $I : \mathcal{G}^N \times \mathcal{N} \to \mathbb{R}$ is said to be *k-monotone positive* if, for any positive integer k and k-order monotonic game $v \in \mathcal{G}^N$,

$$I(v, S) \geq 0 \; \forall S \in \mathcal{N} \text{ whenever } |S| \leq k.$$

As discussed in Subsection 2.2, in a k-monotone game, it seems sensible to consider that there are necessarily complementarity effects among players in coalitions containing (at most) k players. This axiom then simply states that these effects should be represented as positive interactions.

The (Shapley-type) interaction index, as similar to the Shapley value, can be represented via the Möbius transform Δ^v and the multilinear extension $\bar{v}$ of $v \in \mathcal{G}^N$ as follows [18]:

$$I(v, S) = \sum_{T \supseteq S} \frac{1}{t - s + 1} \Delta^v(T).$$

$$I(v, S) = \int_0^1 \frac{\partial^{|S|}}{\partial x_{\sigma(1)} \cdots \partial x_{\sigma(|S|)}} \bar{v}(t, t, \cdots, t)\, dP(t),$$

where σ is a permutation on N such that $S = \{\sigma(1), \cdots, \sigma(|S|)\}$ and $P(t) = t$ for any $t \in [0, 1]$.

Note (other interaction indices): Another Shapley-type interaction index I_{ch} called *chaining interaction index* and the Banzhaf-type interaction index I_B have been proposed and characterized axiomatically (see, e.g., [18,27,49]).

The chaining interaction index $I_{ch} : \mathcal{G}^N \times \mathcal{N} \to \mathbb{R}$ is defined by

$$I_{ch}(v, S) := \sum_{T \subseteq N \setminus S} \frac{s(n - s - t)!(s + t - 1)!}{n!} \Delta_S v(T)$$

for any $v \in \mathcal{G}^N$ and any $S \in \mathcal{N}$, and also has the following representations:

$$I_{ch}(v, S) = \sum_{T \supseteq S} \frac{s}{t} \Delta^v(T).$$

$$I_{ch}(v, S) = \int_0^1 \frac{\partial^{|S|}}{\partial x_{\sigma(1)} \cdots \partial x_{\sigma(|S|)}} \bar{v}(t, t, \cdots, t)\, dP_{ch}(t),$$

where σ is a permutation on N such that $S = \{\sigma(1), \cdots, \sigma(|S|)\}$ and $P_{ch}(t) = t^s \mathbf{1}_{]0,1]}$ for any $t \in [0, 1]$.

The Banzhaf-type interaction index $I_B : \mathcal{G}^N \times \mathcal{N} \to \mathbb{R}$ is defined by

$$I_B(v, S) := \sum_{T \subseteq N \setminus S} \frac{1}{2^{n-s}} \Delta_S v(T)$$

for any $v \in \mathcal{G}^N$ and any $S \in \mathcal{N}$, and also has the following representations:

$$I_B(v, S) = \sum_{T \supseteq S} \frac{1}{2^{t-s}} \Delta^v(T).$$

$$I_B(v, S) = \int_0^1 \frac{\partial^{|S|}}{\partial x_{\sigma(1)} \cdots \partial x_{\sigma(|S|)}} \bar{v}(t, t, \cdots, t)\, dP_B(t),$$

where σ is a permutation on N such that $S = \{\sigma(1), \cdots, \sigma(|S|)\}$ and $P_B(t) = \mathbf{1}_{[0.5,1]}$ for any $t \in [0, 1]$.

3 Bi-cooperative Game

To date, there have been some attempts to define more general concept in cooperative game theory. Aubin [1] has proposed the concept of *generalized coalition* as a function $c : N \to [-1, 1]$ which associates each player i with his/her level of participation $c(i) \in [-1, 1]$. A positive level is interpreted as attraction of the player i for the coalition, and a negative level as repulsion. Later, the concept of bi-cooperative game has been introduced by Bilbao et al. [5] as a generalization of classical cooperative games, where each player can participate positively to the game (defender), negatively (defeater), or do not participate (abstentionist). In a voting situation (simple games), they coincide with ternary voting games, on the set of all *signed coalitions* given by $\{c : N \to \{-1, 0, 1\}\}$, of Felsenthal and Machover [15], where each voter can vote in favor (1), against (-1) or abstain (0). Labreuche and Grabisch [43] give the following example:

Example 1. A set N of farmers raise three kinds of plants called A, B and C (for instance colza, grass and reed) in a given area. Plant A (defeater) needs a lot of pesticide and chemical fertilizers so that it pollutes a lot the local river. Plant B (abstentionist) needs no special treatment and thus no pollution is caused by this plant. Plant C (defender) helps in reducing the pollution since it absorbs some chemicals. The Governor of this area wants to determine the tax for each farmer on the basis of the impact of the farming on the river pollution rate. The bi-cooperative game $v(S, T)$ measures the pollution rate in the river compared to the time when there were only meadows in the area, when farmers S raise plant C, farmers T raise plant A and farmers $N \setminus (S \cup T)$ raise plant B.

3.1 Definitions and Several Representations of Bi-cooperative Games

We will denote $\mathcal{P}(N) := 2^N$ and $\mathcal{Q}(N) := \{(A_1, A_2) \in \mathcal{P}(N) \times \mathcal{P}(N) \mid A_1 \cap A_2 = \emptyset\}$. When equipped with the following order: for $(A_1, A_2), (B_1, B_2) \in \mathcal{Q}(N)$

$$(A_1, A_2) \sqsubseteq (B_1, B_2) \text{ iff } A_1 \subseteq B_1 \text{ and } A_2 \supseteq B_2,$$

$(\mathcal{Q}(N), \sqsubseteq)$ becomes a lattice, which will be defined in Definition 42. The binary operators $\sqcup$ (sup) and $\sqcap$ (inf) are given by

$$(A_1, A_2) \sqcup (B_1, B_2) = (A_1 \cup B_1, A_2 \cap B_2),$$
$$(A_1, A_2) \sqcap (B_1, B_2) = (A_1 \cap B_1, A_2 \cup B_2).$$

Then the top and bottom are respectively $(N, \emptyset)$ and $(\emptyset, N)$.

Definition 19 (irreducible elements [14]). Let $(L, \leq, \vee, \wedge, \top, \bot)$ be a lattice, where $\vee, \wedge, \top, \bot$ denotes sup, inf, the top and bottom element, respectively. An element $x \in L$ is said to be $\vee$*-irreducible* if $x \neq \bot$ and $x = a \vee b$ implies $x = a$ or $x = b$, $\forall a, b \in L$.

Proposition 3 ([24]). *The* $\sqcup$*-irreducible elements of* $\mathcal{Q}(N)$ *are* $(\emptyset, N \setminus i)$ *and* $(i, N \setminus i)$*, for all* $i \in N$*. Moreover, for any* $(A_1, A_2) \in \mathcal{Q}(N)$*,*

$$(A_1, A_2) = \bigsqcup_{i \in A_1} (i, N \setminus i) \ \sqcup \bigsqcup_{j \in N \setminus (A_1 \cup A_2)} (\emptyset, N \setminus j). \tag{3}$$

Eq. (3) is called the *minimal decomposition* of (A_1, A_2) [23].

Here, $\sqcup$-irreducible elements permit to define layer *in $Q(N)$ as follows [23]: $(\emptyset, N)$ is the bottom layer (layer 0) (the black square in Fig. 1), the set of all $\sqcup$-irreducible elements forms layer 1 (black circles in Fig. 1), and layer k, for $k = 2, \ldots n$, consists of all elements whose minimal decomposition contains exactly k $\sqcup$-irreducible elements. In other words, layer k consists of all elements $(A_1, A_2) \in Q(N)$ such that $|A_2{}^c| = k$, for $k = 2, \ldots, n$. On the other hand, let us consider the Boolean lattice $(\mathcal{P}(N), \subseteq, \cup, \cap, N, \emptyset)$. Then, the empty set is the bottom layer; all singletons are $\cup$-irreducible elements, (i.e., in layer 1); the set of all $A \in \mathcal{P}(N)$ whose cardinality is k, for $k = 2, \ldots, n$, forms layer k.*

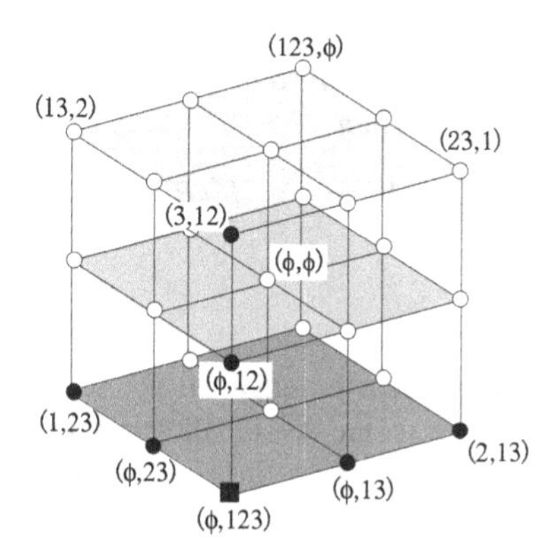

Fig. 1. The lattice $Q(123)$: the element in layer 0 is indicated by a black square and elements in layer 1 black circles

Definition 20 (bi-cooperative game [3,24]). The function v that assigns to every bi-coalition $(S_1, S_2) \in Q(N)$ its value or worth $v(S_1, S_2)$ is commonly referred to as the *bi-characteristic function*. It is always assumed that $v(\emptyset, \emptyset) = 0$. A triplet $(N, v, Q(N))$ consisting of a player set N, a bi-characteristic function v, and a lattice $Q(N)$ constitutes a *bi-cooperative game* or *bi-coalitional game*. (We often identify $(N, v, Q(N))$ with v).

Definition 21 (the Möbius transform of bi-cooperative game). To any bi-cooperative game $v : Q(N) \to \mathbb{R}$, another function $m : Q(N) \to \mathbb{R}$ can be associated by

$$\Delta^v(A_1, A_2) := \sum_{\substack{B_1 \subseteq A_1 \\ A_2 \subseteq B_2 \subseteq N \setminus A_1}} (-1)^{|A_1 \setminus B_1| + |B_2 \setminus A_2|} \, v(B_1, B_2)$$

for $(A_1, A_2) \in Q(N)$. This correspondence proves to be one-to-one, since conversely

$$v(A_1, A_2) = \sum_{(B_1, B_2) \sqsubseteq (A_1, A_2)} \Delta^v(B_1, B_2) \tag{4}$$

for all $(A_1, A_2) \in Q(N)$. The validity of Eq. (4) is proved by Grabisch and Labreuche [23] who call $\Delta^v : Q(N) \to \mathbb{R}$ the *Möbius transform* of v.

Fujimoto and Murofushi [20] have introduced another equivalent representation, the *bipolar Möbius transform*, of a bi-cooperative game as follows:

Definition 22 (the bipolar Möbius transform). To any bi-cooperative game $v : \mathcal{Q}(N) \to \mathbb{R}$, another function $b : \mathcal{Q}(N) \to \mathbb{R}$ can be associated by

$$b^v(A_1, A_2) := \sum_{\substack{B_1 \subseteq A_1 \\ B_2 \subseteq A_2}} (-1)^{|A_1 \setminus B_1| + |A_2 \setminus B_2|} \, v(B_1, B_2) \quad (5)$$

$$= \sum_{(\emptyset, A_2) \sqsubseteq (B_1, B_2) \sqsubseteq (A_1, \emptyset)} (-1)^{|A_1 \setminus B_1| + |A_2 \setminus B_2|} \, v(B_1, B_2)$$

for $(A_1, A_2) \in \mathcal{Q}(N)$. Then, the function defined by Eq. (5) is called the *bipolar Möbius transform* of v.

Proposition 4 ([20]). *Let $v : \mathcal{Q}(N) \to \mathbb{R}$ be a bi-cooperative game, and $b^v : \mathcal{Q}(N) \to \mathbb{R}$ the bipolar Möbius transform of v. Then,*

$$v(A_1, A_2) = \sum_{\substack{B_1 \subseteq A_1 \\ B_2 \subseteq A_2}} b^v(B_1, B_2)$$

for any $(A_1, A_2) \in \mathcal{Q}(N)$.

Definition 23 (piecewise multi linear extension). The set of all bi-cooperative games $\{v : \mathcal{Q}(N) \to \mathbb{R}\}$ is isomorphic to the set of all *ternary pseudo-Boolean functions* $\{f : \{-1, 0, 1\}^N \to \mathbb{R}\}$. Indeed, there exists the isomorphism $\varphi : \mathcal{Q}(N) \to \{-1, 0, 1\}^N$ such that $\varphi(A_1, A_2) = \chi_{(A_1, A_2)}$ for any $(A_1, A_2) \in \mathcal{Q}(N)$, where $\chi_{(A_1, A_2)}$ denotes the characteristic vector of (A_1, A_2), which is the vector of $\{-1, 0, 1\}^N$ whose i-th element is 1 if $i \in A_1$, -1 if $i \in A_2$, and 0 otherwise. Then, for any bi-cooperative game $v : \mathcal{Q}(N) \to \mathbb{R}$ there exists a ternary pseudo-Boolean function f_v (i.e., $f_v : \{-1, 0, 1\}^N \to \mathbb{R}$) corresponding to v. Now, we introduce an equivalent representation, by using a ternary pseudo-Boolean function, of v as follows:

$$f_v(\boldsymbol{x}) = \sum_{(S_1, S_2) \in \mathcal{Q}(N)} b^v(S_1, S_2) \left(\prod_{i \in S_1} x_i^+ \cdot \prod_{j \in S_2} x_j^- \right) \quad (6)$$

for $\boldsymbol{x} \in \{-1, 0, 1\}^N$, where $x^+ = \max\{x, 0\}$ and $x^- = 0 - \min\{x, 0\}$. This correspondence is represented as

$$f_v(\chi_{(A_1, A_2)}) = \sum_{\substack{B_1 \subseteq A_1 \\ B_2 \subseteq A_2}} b^v(B_1, B_2) = v(A_1, A_2) \quad \forall (A_1, A_2) \in \mathcal{Q}(N).$$

Here, Eq. (6) leads to the *piecewise multilinear extension* $g_v : [-1, 1]^N \to \mathbb{R}$, of the ternary pseudo-Boolean function $f_v : \{-1, 0, 1\}^N \to \mathbb{R}$ corresponding to the bi-cooperative game $v : \mathcal{Q}(N) \to \mathbb{R}$, defined by

$$g_v(\boldsymbol{x}) := \sum_{(S_1, S_2) \in \mathcal{Q}(N)} b^v(S_1, S_2) \left(\prod_{i \in S_1} x_i^+ \cdot \prod_{j \in S_2} x_j^- \right)$$

for $\boldsymbol{x} \in [-1, 1]^N$.

3.2 Monotonicity and Derivatives

As seen in Section 2.2 and 2.4, the definition of the derivative is the key concept for the interaction index. Grabisch and Labreuche [23] extended the notion of *discrete derivative* of ordinary cooperative games to that of bi-cooperative games.

Definition 24 **((T_1, T_2)-derivative of bi-cooperative game).** Let $(N, v, Q(N))$ be a bi-cooperative game. For $(T_1, T_2) \in Q(N)$, the (T_1, T_2)-*derivative* at a point $(S_1, S_2 \cup T_2) \in Q(N)$, where $(S_1, S_2) \in Q(N \setminus (T_1 \cup T_2))$, is denoted as $\Delta_{(T_1,T_2)} v(S_1, S_2 \cup T_2)$ and defined by

$$\begin{aligned}\Delta_{(T_1,T_2)} v(S_1, S_2 \cup T_2) &:= \sum_{\substack{L_1 \subseteq T_1 \\ L_2 \subseteq T_2}} (-1)^{|T_1 \setminus L_1| + |L_2|} \, v(S_1 \cup L_1, S_2 \cup L_2) \\ &= \sum_{(S_1, S_2 \cup T_2) \sqsubseteq (A_1, A_2) \sqsubseteq (S_1 \cup T_1, S_2)} (-1)^{|A_1 \setminus S_1| + |A_2 \setminus S_2|} \, v(A_1, A_2).\end{aligned} \tag{7}$$

The formula (7) is led by the following recursive relations [24]:

$$\begin{aligned}\Delta_{(i,\emptyset)} v(S_1, S_2) &:= v(S_1 \cup i, S_2) - v(S_1, S_2), \\ \Delta_{(\emptyset,j)} v(S_1, S_2 \cup j) &:= v(S_1, S_2) - v(S_1, S_2 \cup j), \\ \Delta_{(T_1,T_2)} v(S_1, S_2 \cup T_2) &:= \Delta_{(i,\emptyset)} \left(\Delta_{(T_1 \setminus i, T_2)} v(S_1, S_2 \cup T_2)\right) \\ &= \Delta_{(\emptyset,j)} \left(\Delta_{(T_1, T_2 \setminus j)} v(S_1, (S_2 \cup j) \cup (T_2 \setminus j))\right),\end{aligned}$$

where $i \in T_1, j \in T_2$.

Example 2. Let us consider the $(12, 3)$-derivative at $(\emptyset, 3)$. Then, $\Delta_{(12,3)} v(\emptyset, 3)$ is represented by

$$\Delta_{(12,3)} v(\emptyset, 3) = \Delta_{(1,3)} v(2, 3) - \Delta_{(1,3)} v(\emptyset, 3).$$

The derivatives $\Delta_{(1,3)} v(2, 3)$ and $\Delta_{(1,3)} v(\emptyset, 3)$ are represented by

$$\Delta_{(1,3)} v(2, 3) = \Delta_{(\emptyset,3)} v(12, 3) - \Delta_{(\emptyset,3)} v(2, 3) \text{ and } \Delta_{(1,3)} v(\emptyset, 3) = \Delta_{(\emptyset,3)} v(1, 3) - \Delta_{(\emptyset,3)} v(\emptyset, 3),$$

respectively. The derivatives $\Delta_{(\emptyset,3)} v(12, 3)$, $\Delta_{(\emptyset,3)} v(2, 3)$, $\Delta_{(\emptyset,3)} v(1, 3)$ and $\Delta_{(\emptyset,3)} v(\emptyset, 3)$ are represented by

$$\Delta_{(\emptyset,3)} v(12, 3) = v(12, \emptyset) - v(12, 3), \quad \Delta_{(\emptyset,3)} v(2, 3) = v(2, \emptyset) - v(2, 3),$$

$$\Delta_{(\emptyset,3)} v(1, 3) = v(1, \emptyset) - v(1, 3), \text{ and } \Delta_{(\emptyset,3)} v(\emptyset, 3) = v(\emptyset, \emptyset) - v(\emptyset, 3),$$

respectively. Inversely, first, consider the first order derivatives $\Delta_{(\emptyset,3)}$ at $(\emptyset, 3)$, $(1, 3)$, $(2, 3)$, and $(12, 3)$. These derivatives correspond to thin arrow lines in Fig. 2, respectively. Second, the second order derivatives $\Delta_{(1,3)}$ at $(\emptyset, 3)$ and $(2, 3)$ correspond to thick black arrow lines in Fig. 2, which represent the differences between the first order derivatives represented by thin arrow lines. Finally, the $(12, 3)$-derivative at $(\emptyset, 3)$ corresponds to the thick gray arrow line in Fig. 2.

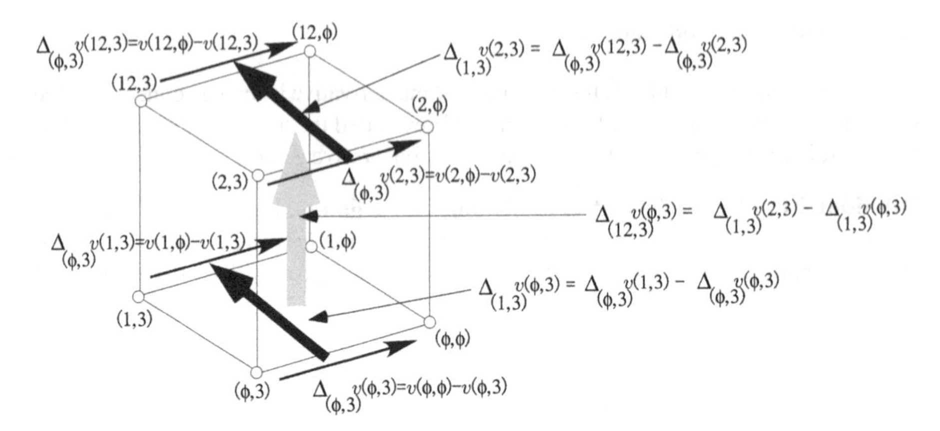

Fig. 2. The $(12, 3)$-derivative at $(\emptyset, 3)$

Proposition 5 ([20,23]). *For* $(T_1, T_2) \in \mathcal{Q}(N)$,

$$\varDelta_{(T_1,T_2)}v(S_1, S_2 \cup T_2) = \sum_{(T_1, N\setminus(T_1\cup T_2))\sqsubseteq(A_1,A_2)\sqsubseteq(S_1\cup T_1, S_2)} \varDelta^v(A_1, A_2)$$
$$= (-1)^{|T_2|} \sum_{\substack{L_1\subseteq S_1\\ L_2\subseteq S_2}} b^v(L_1 \cup T_1, L_2 \cup T_2)$$

for any $(S_1, S_2) \in \mathcal{Q}(N \setminus (T_1 \cup T_2))$. *Thus,*

$$\varDelta^v(T_1, T_2) = \varDelta_{(T_1,T_2)}v(\emptyset, N \setminus T_1) \quad \forall (T_1, T_2) \in \mathcal{Q}(N).$$

Proposition 6 ([20]). *Let* $v : \mathcal{Q}(N) \to \mathbb{R}$ *be a bi-cooperative game and* $g_v : [-1, 1]^N \to \mathbb{R}$ *the piecewise multilinear extension of the ternary pseudo-Boolean function corresponding to* v, *and* $(T_1, T_2) := (\{t_1(1), \ldots, t_1(p)\}, \{t_2(1), \ldots, t_2(q)\}) \in \mathcal{Q}(N)$, *where* $|T_1| = p$ *and* $|T_2| = q$. *Then,*

$$\varDelta_{(T_1,T_2)}v(S_1, S_2 \cup T_2) = \left.\frac{\partial^{(p+q)}}{\partial x_{t_1(1)} \cdots \partial x_{t_1(p)} \partial x_{t_2(1)} \cdots \partial x_{t_2(q)}} g_v(\boldsymbol{x})\right|_{\boldsymbol{x}=\chi_{(S_1\cup T_1, S_2\cup T_2)}}$$

for all $(S_1, S_2) \in \mathcal{Q}(N \setminus (T_1 \cup T_2))$, *where* $\chi_{(S_1\cup T_1,S_2\cup T_2)} \in [-1, 1]^N$ *is the characteristic vector of* $(S_1 \cup T_1, S_2 \cup T_2) \in \mathcal{Q}(N)$. *It should be noticed that* $\dfrac{\partial}{\partial x_i} g_v$ *cannot be defined on* $\{\boldsymbol{x} \in [-1, 1]^N \mid x_i = 0\}$.

Labreuche and Grabisch [42] have proposed the notion of *k-monotonicity* in bi-cooperative games as a bipolar extension of that in ordinary cooperative games.

Definition 25 (*k*-monotonic bi-cooperative game [24,42]). Given an integer $k \geq 2$, a bi-cooperative game $v : \mathcal{Q}(N) \to \mathbb{R}$ is said to be *k-order monotone* (for short, *k-monotone*) if and only if, for any (at most) k bi-coalitions $S_1, \cdots, S_k$, we have

$$v\left(\bigsqcup_{i=1}^{k} S_i\right) \geq \sum_{\substack{J\subseteq\{1,\cdots,k\}\\ J\neq\emptyset}} (-1)^{|J|+1}\, v\left(\bigsqcap_{j\in J} S_j\right). \tag{8}$$

It is easy to verify that k-monotonicity ($k > 2$) implies l-monotonicity for all integer $2 \leq l \leq k$. By extension, 1-monotonicity (which does not correspond to $k = 1$ in Eq. (8)) is defined as standard monotonicity, i.e.,

$$v(S) \leq v(T) \quad \text{whenever} \quad (\emptyset, N) \sqsubseteq S \sqsubseteq T \sqsubseteq (N, \emptyset).$$

The notion of k-monotonicity of bi-cooperative games can be characterized via discrete derivatives in a similar way to ordinary cases.

Proposition 7 ([18]). *Let $v : Q(N) \to \mathbb{R}$ be a bi-cooperative game and k a positive integer. Then v is k order monotone if and only if*

$$\Delta_{(T_1,T_2)} v(S_1, S_2 \cup T_2) \geq 0$$

for any $(T_1, T_2) \in Q(N)$ such as $1 \leq |T_1 \cup T_2| \leq k$ and any $(S_1, S_2) \in Q(N \setminus (T_1 \cup T_2))$.

3.3 Value and Interaction Index in Bi-cooperative Games

In this subsection, we introduce the notion of value and power index for bi-cooperative games and ternary simple games, following Labreuche and Grabisch [43], in the spirit of what was done by Shapley [59] for cooperative games, and by Shapley and Shubik [60] for simple games. In ordinary cooperative games, an allocation rule (pre-imputation) is a vector $\boldsymbol{x} \in \mathbb{R}^N$ which represents the share of the total worth of the game $v(N)$ among the players, assuming that all players have decided to join the grand coalition N. For bi-cooperative games, the situation differs since apart from not participating to the game, each player has two possible actions, namely to play in the defender or the defeater part, while he/she has only one in classical (ordinary cooperative game) case. In order to generalize the notion of imputation, the concept of *reference action* or *level* has been introduced. The reference action is the action such that if all players do this action, then the outcome of the game is 0. For ordinary games, the reference action is to "not participate" since $v(\emptyset) = 0$. For bi-cooperative games, it is also the non participation since $v(\emptyset, \emptyset) = 0$. An imputation is defined for each possible action (except the reference one) of a player with respect to the reference action, that is, it represents a kind of average contribution of the player for a given action, compared to the reference action. For bi-cooperative games, the possible actions are: to play in the defender part, or to play in the defeater part. For preserving the meaning of "contribution" (which has a positive sense) and for compatibility with previous works, it requires to consider two values ϕ^+ (defender part) and ϕ^- (defeater part): ϕ^+ is the contribution of "playing in the defender part" instead of "doing nothing", and ϕ^- is the contribution of "doing nothing" instead of "playing in the defeater part". Then, an overall contribution ϕ can be defined as $\phi = \phi^+ + \phi^-$.

Labreuche and Grabisch [43] have proposed the following value for bi-cooperative games, axiomatically.

Definition 26 (value of bi-cooperative game [43]). The *(Shapley-type) value* ϕ^B of a bi-cooperative game $v : \mathcal{Q}(N) \to \mathbb{R}$ is given by

$$\phi_i^B(v) := \sum_{S \subseteq N \setminus i} \frac{s!\,(n-s-1)!}{n!} \left[v(S \cup i, N \setminus (S \cup i)) - v(S, N \setminus S) \right]$$

$$= \sum_{(S,T) \sqsupseteq (\emptyset, N \setminus i)} \frac{1}{n-t} \Delta^v(S,T)$$

for every $i \in N$, where $\phi_i^B(v)$ is the i-th component of $\phi^B(v) \in \mathbb{R}^N$. The value ϕ^B is decomposable into two values ϕ^+ (defender part) and ϕ^- (defeater part), i.e., $\phi^B = \phi^+ + \phi^-$, as follows:

$$\phi_i^+(v) = \sum_{S \subseteq N \setminus i} \frac{s!\,(n-s-1)!}{n!} \left[v(S \cup i, N \setminus (S \cup i)) - v(S, N \setminus (S \cup i)) \right],$$

$$\phi_i^-(v) = \sum_{S \subseteq N \setminus i} \frac{s!\,(n-s-1)!}{n!} \left[v(S, N \setminus (S \cup i)) - v(S, N \setminus S) \right].$$

The (Shapley-type) value ϕ^B for bi-cooperative game is a kind of generalization of the Shapley value of ordinary cooperative games in the following sense:

If a bi-cooperative game $v^B : \mathcal{Q}(N) \to \mathbb{R}$ is given via some ordinary cooperative game $v : 2^N \to \mathbb{R}$ as

$$v^B(S,T) = \begin{cases} v(S) & \text{if } T = \emptyset, \\ 0 & \text{otherwise,} \end{cases}$$

for any $(S,T) \in \mathcal{Q}(N)$, then

$$\phi^B(v^B) = \phi^+(v^B) = \phi(v), \quad \phi^-(v^B) = 0,$$

where $\phi(v)$ is the Shapley value of ordinary game v.

The *(Shapley-type) interaction index for bi-cooperative games* has been introduced by Grabisch and Labreuche [24,26] and characterized axiomatically by Lange and Grabisch [45], by analogy with that for ordinary cooperative games.

Definition 27 (interaction index for bi-cooperative game). The *(Shapley-type) interaction index* $I^B(v,(S,T))$ with respect to $(S,T) \in \mathcal{Q}(N)$ of a bi-cooperative game $v : \mathcal{Q}(N) \to \mathbb{R}$ is defined by

$$I^B(v,(S,T)) := \sum_{U \subseteq N \setminus (S \cup T)} \frac{(n-s-t-u)!\,u!}{(n-s-t+1)!} \Delta_{(S,T)}(U, N \setminus (S \cup U))$$

$$= \sum_{\substack{(A,B) \sqsupseteq (S, N \setminus (S \cup T)) \\ (A,B) \sqsubseteq (N \setminus T, \emptyset)}} \frac{1}{n-s-t-b+1} \Delta^v(A,B).$$

Kojadinovic [41] has proposed another interaction index for bi-cooperative games in the context of aggregation by the bipolar Choquet integral [25], however his solution is not completely axiomatized.

Definition 28 (Kojadinovic's interaction index). The *(Kojadinovic-type) interaction index* $I^K(v,(S_1,S_2))$ with respect to $(S_1,S_2) \in \mathcal{Q}(N)$ of a bi-cooperative game $v : \mathcal{Q}(N) \to \mathbb{R}$ is defined by

$$I^K(v,(S_1,S_2)) := \sum_{(T_1,T_2)\in\mathcal{Q}(N\setminus S)} \frac{1}{2^t}\frac{(n-s-t+1)!\,t!}{(n-s+1)!}\Delta_{(S_1,S_2)}(T_1,T_2\cup S_2),$$

where $T := T_1 \cup T_2$ and $S := S_1 \cup S_2$.

Note (Ternary voting games and its values) : Bi-cooperative games are a generalization of the notion of *ternary voting game* which has been proposed by Felsenthal and Machover [15]. In a play of a ternary voting game, each player can choose between voting *in favor* (yes), *against* (no), or *abstaining*. Formally, a ternary voting game is a function $v : \mathcal{Q}(N) \to \{-1, 1\}$, where $v(F, A)$ represents the result of the vote (1 if the bill is passed, -1 if it is defeated) when voters in F vote in favor, voters in A vote against, and the remaining voters abstain. Then, obviously, it should be satisfied that

$$v(\emptyset,N) = -1, \quad v(N,\emptyset) = 1, \text{ and } \quad (F_1,A_1) \sqsubseteq (F_2,A_2) \implies v(F_1,A_1) \leq v(F_2,A_2).$$

Definition 29 (ternary roll-call and pivot). A *ternary roll-call* R is a triplet $R = (\sigma_R, F_R, A_R)$ consisting of a permutation σ_R on N, a coalition F_R which contains all voters that are in favor of the bill, and a coalition A_R which contains all voters that are against the bill. Ternary roll-calls are interpreted as follows. The voters are called in order given by $\sigma_R : \sigma_R(1), \ldots, \sigma_R(n)$. When a voter i is called, he/she tells his/her opinion, that is to say *in favor* if $i \in F_R$, *against* if $i \in A_R$, or abstention otherwise. The set of all ternary roll-calls on N is denoted by $\mathcal{T}_N$, whose cardinality is $3^n \cdot n!$.

A *pivot* $Piv(v,R)$ for a ternary voting game $(N, v, \mathcal{Q}(N))$ and a ternary roll-call $R = (\sigma_R, F_R, A_R)$ is the player i, represented by $i = \sigma_R(m)$ for some $m \in \{1, \ldots, n\}$, satisfying the following two conditions:

1. $v(F_R^{m-1}, A_R^{m-1}) \neq v(F_R^m, A_R^m)$ if $m \neq 1$.
2. $v(F_R^m, A_R^m) = v(F_R^k, A_R^k)$ for any $k > m$,

where $F_R^k := \bigcup_{j\leq k}\{\sigma_R(j)\}\cap F_R$ and $A_R^k := \bigcup_{j\leq k}\{\sigma_R(j)\}\cap A_R$. That is, $Piv(v,R)$ is decisive in the result of the vote.

Felsenthal and Machover [15] have proposed a voting power index $\phi^{ter}(v)$ for a ternary voting game $(N, v, \mathcal{Q}(N))$, which is a generalization of the Shapley-Shubik power index, as follows:

$$\phi_i^{ter}(v) := \frac{|\{R \in \mathcal{T}_N \mid i = Piv(v,R)\}|}{|\mathcal{T}_N|}.$$

However, this solution is not completely axiomatized.

4 Cooperative Game and Network

In ordinary cooperative game theory, it is implicitly assumed that all coalitions of N can be formed, this is in general not the case. In order for players to be able to coordinate their actions, they have to be able to communicate. The bilateral communication channels between players in N are described by a *communication network*. Such a network can be represented by an *undirected graph* $G = (N, E)$, which has the set of players as its *nodes* $S \subseteq N$ and in which those players are connected by the set of *edges/links* $E \subseteq \{ij \mid i, j \in N, i \neq j\}$, i.e., players i and j can communicate (directly) with each other if $ij \in E$. To avoid cumbersome notations, we often omit braces for a graph $G = (N, E)$, we denote $E - ij := E \setminus ij$ for any $ij \in E$, $E + ij := E \cup ij$ for any $ij \notin E$, and $G(S) := (S, E(S))$ for any $S \subseteq N$, where $E(S) := \{ij \in E \mid i, j \in S\}$. Then, $G(S)$ is called the subgraph induced from the underlying graph G and the subset S of N.

The number of the nodes adjacent to a node $i \in N$ is said to be the *degree* of i in $G = (N, E)$ and denoted by $a^G(i)$. A sequence of different nodes $(i_1, \ldots, i_m)$ is called a *path*, whose length is $m - 1$, between j and k in a graph (N, E) if $j = i_1$, $k = i_m$, and $\{i_l, i_{l+1}\} \in E$ for any $l \in \{1, \ldots, m-1\}$. If there is a path between j and k in an undirected graph G, then we say that j is *reachable* to k in G and denote $j \sim_G k$. A set of nodes $S \subseteq N$ is called *connected* in an undirected graph $G := (N, E)$ if for any $i, j \in S$, $i \neq j$, there exists a path $(i_1, \ldots, i_m)$ between i and j in G satisfying that all nodes of the path are in S, i.e., $i_k \in S$ for any $k \in \{1, \ldots, m\}$. Notice that, by definition, the empty set and all singletons are *connected*. An undirected graph $G = (N, E)$ is said to be connected if N is connected in G. Clearly, the relation $\sim_G$ is an equivalence relation on N. Hence, the notion of reachableness induces a partition $N/E := N/\sim_G$ of N. Then, for any $S \subseteq N$, $C \in S/E(S) = S/\sim_{G(S)}$ is called a *(connected) component* of S. A *geodesic* (also often called a "shortest" path) between two nodes $i, j \in N$ is a path whose length is the minimum among all paths between i and j. The length of a geodesic between two nodes $i, j \in N$ in G, if i and j are reachable, is called their *geodesic distance*. If i and j are unreachable, then the distance between i and j is infinite. A sequence of nodes $(i_1, \ldots, i_m)$ is called a *cycle* if $i_1 = i_m$ and $(i_1, \ldots, i_{m-1})$ is a path. A graph is *cycle free* if it does not contain any cycle. A connected cycle free graph is called *tree*.

Definition 30 (communication situation). The triplet (N, v, E), which reflects a situation consisting of a game $v \in \mathcal{G}^N$ and a communication network (N, E), is called a *communication situation*. We denote the set of all communication situations on N by CS^N.

Definition 31 (feasible coalition). A coalition $S \subseteq N$ is said to be *feasible* in the communication network $G = (N, E)$ if S is connected in G (i.e., $S/E = \{S\}$).

Example 3. Consider the communication situation (N_1, v, E_1) with $N_1 = \{1, 2, \cdots, 7\}$ and $E_1 = \{12, 15, 26, 37, 47, 56\}$ (Fig.3). Then, all the players in $\{1, 2, 6\}$ can communicate with one another, i.e., the coalition $\{1, 2, 6\}$ is feasible. Hence, they can fully coordinate their actions and obtain the value $v(\{1, 2, 6\})$. On the other hand, in the coalition $\{1, 2, 3, 4\}$, players 1 and 2 are reachable, however, both of players 3 and 4 cannot communicate with any other players in $\{1, 2, 3, 4\}$. Then, feasible subcoalitions of

$\{1, 2, 3, 4\}$ are $\{1, 2\}$, $\{3\}$, and $\{4\}$ (i.e., $\{1, 2, 3, 4\}/E_1 = \{12, 3, 4\}$, thus forming the coalition $\{1, 2, 3, 4\}$ is unfeasible). Hence, the value attainable by the players in $\{1, 2, 3, 4\}$ should be $v(1, 2) + v(3) + v(4)$. That is, in general, the value attainable by the players in S under a communication situation (N, v, E) is represented by $\sum_{T \in S/E} v(T)$.

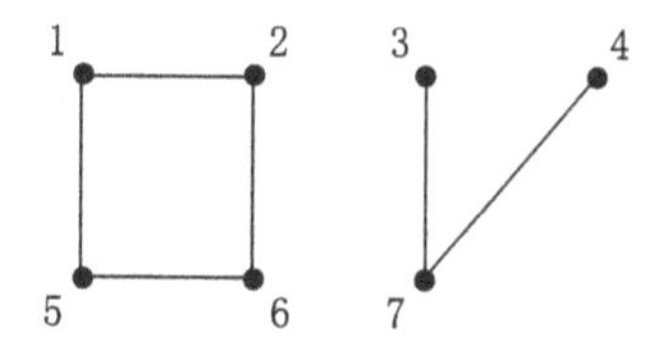

Fig. 3. Communication Network(N_1, E_1)

Definition 32 (network-restricted game [52]). The *network-restricted game* (N, v^E) associated with (N, v, E) is defined by

$$v^E(S) := \sum_{T \in S/E} v(T) \quad \text{for each } S \subseteq N. \tag{9}$$

Note that if (N, E) is the complete graph (i.e., $E = \{ij \mid i, j \in N, i \neq j\}$), the network-restricted game v^E is equal to the original game v.

The network-restricted game evaluates the possible gains from cooperation in a communication situation from the point of view of the players. Next example focuses on the importance of communication channels/links in a communication situation.

Example 4. In the communication network E_1 represented by Fig.3, the value obtainable by the players in the grand coalition N is

$$v^{E_1}(N) = v(\{1, 2, 5, 6\}) + v(\{3, 4, 7\}),$$

since $N/E_1 = \{\{1, 2, 5, 6\}, \{3, 4, 7\}\}$. If for some reason the communication link between players 4 and 7 is lost, the communication network E_1 turns to a new communication network $E_2 = \{12, 15, 26, 37, 56\}$. Then, $N/E_2 = \{\{1, 2, 5, 6\}, \{3, 7\}, \{4\}\}$ and the value obtainable by the players in the grand coalition N turns to

$$v^{E_2}(N) = v(\{1, 2, 5, 6\}) + v(\{3, 7\}) + v(\{4\}).$$

Then $v^{E_1}(N) - v^{E_2}(N)$ can be interpreted as a kind of marginal contribution of the link $47 \in E_1$ to the communication network E_1.

Definition 33 (link game [7]). The *link game* associated with (N, v, E) consisting of a zero-normalized game v (i.e., $v(i) = 0$ for any $i \in N$) is a game on E defined by

$$\gamma^v(H) := v^H(N) = \sum_{T \in N/H} v(T) \quad \text{for each } H \subseteq E.$$

The link game $\gamma^v(H)$ represents the worth of communication network $H \subseteq E$ as the worth of the grand coalition in the communication situation (N, v, H) through the network-restricted game v^H. Note that, for an ordinary game v, the link game γ^v is generally not a game on E since $\gamma^v(\emptyset) = \sum_{T\in N/\emptyset} v(T) = \sum_{i\in N} v(i) \neq 0$.

Example 5 (wighted majory voting game). Consider the weighted majority voting situation $(N, [q : s_1, \ldots, s_n])$ with $N = \{1, 2, 3, 4\}$, $s_1 = 35$, $s_2 = 30$, $s_3 = 25$, $s_4 = 10$, and $q = 51$. So, there are 100 members of parliament who are divided among four political parties labeled 1,2,3, and 4, and decisions are made by majority voting. The parties 1,2,3, and 4 have 35, 30, 25, 10 seats, respectively. This situation can be represented by the game v such that

$$v(S) = \begin{cases} 1 & \text{if } \sum_{i\in S} s_i \geq q = 51, \text{ i.e., win,} \\ 0 & \text{if } \sum_{i\in S} s_i < q = 51, \text{ i.e., lose.} \end{cases}$$

Then, the winning coalitions are $\{1, 2\}$, $\{1, 3\}$, $\{2, 3\}$, $\{1, 2, 3\}$, $\{1, 2, 4\}$, $\{1, 3, 4\}$, $\{2, 3, 4\}$, and $\{1, 2, 3, 4\}$. In general, every coalition $S \subseteq N$ could not been formed, due to ideological and policy differences. Suppose that the party 4 cannot form coalitions with any other parties due to ideological differences; the parties 1 and 3 cannot form the coalition $\{1, 3\}$ due to some policy differences but can form a coalition $\{1, 2, 3\}$ through an intermediary, the party 2. Such a situation can be represented by the graph $G := (N, E)$ as shown in **Fig.**4.

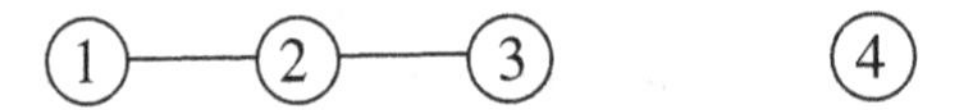

Fig. 4. Relations among political parties

In the network restricted game v^E, the winning coalition are $\{1, 2\}$, $\{2, 3\}$, $\{1, 2, 3\}$, $\{1, 2, 4\}$, $\{2, 3, 4\}$, and $\{1, 2, 3, 4\}$; for instance, $v^E(13) = v(1) + v(3) = 0 + 0 = 0$, i.e., lose, while $v^E(124) = v(12) + v(4) = 1 + 0 = 1$, i.e., win. (However, *feasible* winning coalitions are only $\{1, 2\}$, $\{2, 3\}$, and $\{1, 2, 3\}$).

4.1 Allocation Rule in Communication Situation

In this subsection, we will briefly introduce major two existing values (allocation rules), the *Myerson value* [52] and *the position value* [7], for communication situations.

Definition 34 (the Myerson value [52]). The *Myerson value* for a communication situation (N, v, E) is denoted as $\Psi(N, v, E)$ and defined by

$$\Psi(N, v, E) := \phi(v^E),$$

where $\phi(N, v^E)$ is the Shapley value of (N, v^E). Note that the $\Psi(N, v, E) = \phi(v)$ if (N, E) is the complete graph.

The Myerson value is one of the most famous allocation rules, which assigns to every communication situation (N, v, E) the Shapley value of the network-restricted game (N, v^E) and is characterized as the unique allocation rule satisfying the following two properties/axioms, *component efficiency* and *fairness* (see, Myerson [52]).

Definition 35 (componennt efficiency). For any communication situation $(N, v, E) \in CS^N$, it holds that

$$\sum_{i \in S} \Psi_i(N, v, E) = v(S)$$

for any $S \in N/E$, where $\Psi_i(N, v, E)$ is the i-th component of $\Psi(N, v, E)$.

Definition 36 (fairness). For any communication situation $(N, v, E) \in CS^N$, it holds that

$$\Psi_i(N, v, E) - \Psi_i(N, v, E - ij) = \Psi_j(N, v, E) - \Psi_j(N, v, E - ij)$$

for any $ij \in E$, where $\Psi_i(N, v, E)$ is the i-th component of $\Psi(N, v, E)$.

Component efficiency means that the sum of the players' allocations in a component equal to the worth of the component. Fairness means that the two players connected by a link obtain the same change of allocation if the link is deleted.

Definition 37 (position value [7]). The *position value* for a communication situation (N, v, E) consisting of zero-normalized game v is denoted as $\pi(N, v, E)$ and defined by

$$\pi_i(N, v, E) := \frac{1}{2} \sum_{\substack{e \in E \\ e \ni i}} \phi_e(\gamma^v) \quad \text{for each } i \in N.$$

The Shapley value $\phi_e(\gamma^v)$ of a link $e \in E$ can be interpreted as a kind of *expected marginal contribution* of the link (edge) $e \in E$ to all communication networks containing e. Then, the value is divided equally between the two players at the ends of the considered link $e \in E$. The position value of a given player $i \in N$ is obtained as the sum of all these shares.

Example 6. Consider the communication situation in Example 5. Then, the Shapley value of the underlying game (i.e., the Shapley-Shubik index [60]) is $(\frac{1}{3}, \frac{1}{3}, \frac{1}{3}, 0)$; the Myerson value for the communication situation (N, v, E) (i.e., the Shapley value of the network-restricted game (N, v^E)) is $(\frac{1}{6}, \frac{4}{6}, \frac{1}{6}, 0)$; the position value for the communication situation (N, v, E) is $(\frac{1}{4}, \frac{1}{2}, \frac{1}{4}, 0)$.

4.2 Poset Induced by Communication Network

In this subsection, we consider and introduce a subposet of $(2^N, \subseteq)$ induced by a communication network $G := (N, E)$.

For a communication network $G := (N, E)$, the set of all feasible coalitions in G is denoted as $\mathfrak{F}(G)$, i.e.,

$$\mathfrak{F}(G) := \{S \subseteq N \mid S : \text{connected in } G := (N, E),\ i.e., |S/E| = 1\}.$$

The set $\mathfrak{F}(G)$ together with set inclusion $\subseteq$ as an order on $\mathfrak{F}(G)$ is called the *poset induced by a communication network* G.

Example 7. Let $N = \{1,2,3\}$, $E_a = \{12,13,23\}$, $E_b = \{13,23\}$, and $E_c = \{12\}$. Then the posets induced by communication networks $G_a := (N, E_a)$, $G_b := (N, E_b)$, and $G_c := (N, E_c)$, as shown in (a) – (c) in Fig. 5, are represented as shown in (a) – (c) in Fig. 6, respectively.

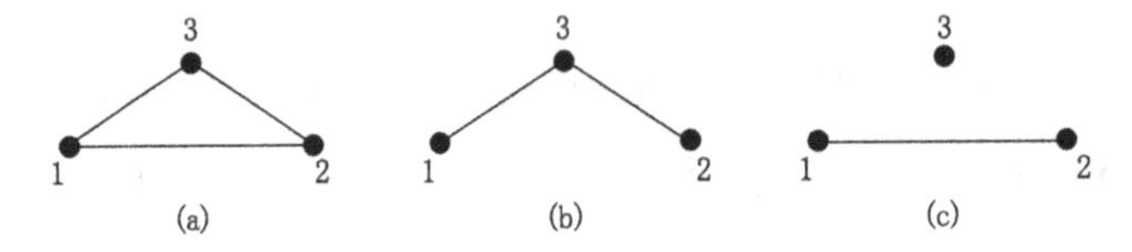

Fig. 5. Communication networks on $N = \{1,2,3\}$

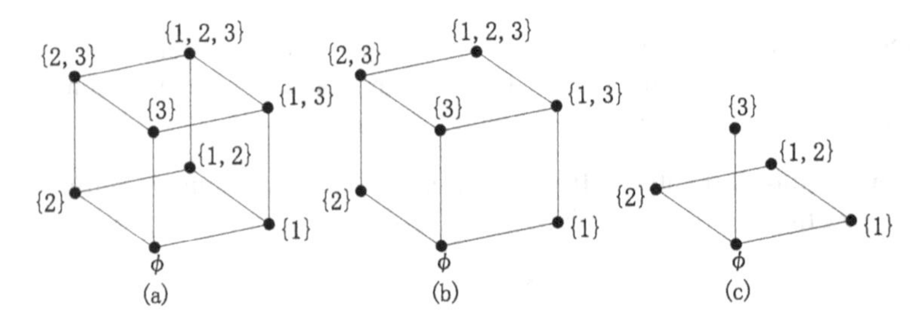

Fig. 6. Posets corresponding to networks in Fig. 5

In a communication situation (N, v, E) consisting of a game (N, v) and a communication network $G := (N, E)$, at least two types of restrictions of $v \in \mathcal{G}^N$ can be considered. One is the network-restricted game v^E defined by Eq. (9). Another is the restriction of $v \in \mathcal{G}^N$, i.e., $v : 2^N \to \mathbb{R}$, to the poset $\mathfrak{F}(G)$ induced by $G = (N, E)$, i.e., $v|_{\mathfrak{F}(G)} : \mathfrak{F}(G) \to \mathbb{R}$. Then it is denoted as $v^{\mathfrak{F}(G)} : \mathfrak{F}(G) \to \mathbb{R}$ and defined by

$$v^{\mathfrak{F}(G)}(S) := v|_{\mathfrak{F}(G)}(S), \ \ i.e., v^{\mathfrak{F}(G)}(S) := v(S), \quad \forall S \in \mathfrak{F}(G). \tag{10}$$

Definition 38 (Möbius transform on poset). Let (N, v, E) be a communication situation consisting of a game (N, v) and a communication network $G := (N, E)$. Then, the *Möbius transform* of $v^{\mathfrak{F}(G)} : \mathfrak{F}(G) \to \mathbb{R}$ on the poset $(\mathfrak{F}(G), \subseteq)$ is denoted by $\Delta^{(N,v,E)}$ and defined through the following equation:

$$v^{\mathfrak{F}(G)}(S) = \sum_{\substack{T \in \mathfrak{F}(G) \\ T \subseteq S}} \Delta^{(N,v,E)}(T) \quad \forall S \in \mathfrak{F}(G).$$

Conversely, $\varDelta^{(N,v,E)}$ is explicitly represented by

$$\varDelta^{(N,v,E)}(S) = \sum_{\substack{T \in \mathfrak{F}(G) \\ T \subseteq S}} (-1)^{|S \setminus T|} v^{\mathfrak{F}(G)}(T) \quad \forall S \in \mathfrak{F}(G).$$

Proposition 8. *Let (N, v, E) be a communication situation consisting of a game (N, v) and a communication network $G := (N, E)$, (N, v^E) denote the network-restricted game defined by Eq. (9), and $v^{\mathfrak{F}(G)}$ the restriction of v to $\mathfrak{F}(G)$ defined by Eq. (10). Then,*

$$\varDelta^{(N,v,E)}(S) = \varDelta^{v^E}(S) \quad \forall S \in \mathfrak{F}(G),$$
$$\varDelta^{v^E}(S) = 0 \quad \forall S \in 2^N \setminus \mathfrak{F}(G).$$

Moreover, the Myerson value $\Psi(N, v, E)$ can be represented as

$$\Psi_i(N, v, E) = \sum_{i \in S \in \mathfrak{F}(G)} \frac{1}{s} \varDelta^{(N,v,E)}(S).$$

4.3 Harsanyi Power Solution for Communication Situation

In this subsection, we introduce a class of allocation rules, *Harsanyi power solutions*, to which many existing allocation rules for communication situations belong. Briks et al. [9] have introduced the concept of *Harsanyi power solution* for communication situations, which is based on Harsanyi solutions for TU-games. The concept of Harsanyi solution is proposed as a class of solutions for TU-games in Vasil'ev [63,64] (see also Derks et al. [12], where a Harsanyi solution is called a sharing value). The idea behind a Harsanyi solution is that it distributes the Harsanyi dividends over the players in the corresponding coalitions according to a chosen sharing system which assigns to every coalition S a sharing vector specifying for every player in S its share in the dividend $\varDelta^v(S)$ of S. The payoff to each player $i \in N$ is thus equal to the sum of its shares in the dividends of all coalitions of which he is a member. A famous Harsanyi solution is the Shapley value.

Now, we consider the case $N = \{1, 2\}$, the Shapley value $\phi_1(N, v)$ of player 1 in a game (N, v) is obtained as

$$\phi_1(N, v) = \frac{1}{1}\varDelta^v(1) + \frac{1}{2}\varDelta^v(12).$$

This expression, as an allocation rule of *Harsanyi dividends* (i.e., the Möbius transform), has the following (at least two) interpretations:

Interpretation 1 (Egalitarian allocation) : *The Shapley value distributes the dividend of any coalition S equally among the players in S, i.e., $\frac{1}{s}\varDelta^v(S)$, (so players outside S do not share in the dividend of S).*

Interpretation 2 (Allocation based on coalition forming process) : *We consider a process to form the coalition* $\{1,2\}$. *Then, there are two shortest paths from* $\emptyset$ *to* $\{1,2\}$ *in Fig. 7* (a). *One is the path* $\emptyset \to \{1\} \to \{1,2\}$; *another is the path* $\emptyset \to \{2\} \to \{1,2\}$. *The path* $\emptyset \to \{1\} \to \{1,2\}$ *can be interpreted as follows: Player 1 makes an offer to player 2 for forming the coalition* $\{1,2\}$. *Player 2 accepts the offer and adds to the coalition* $\{1\}$ *to form the new coalition* $\{1,2\}$. *Among these two paths, the only path that passes through* $\{1\}$, *i.e., the player 1 plays a role of initiator in forming* $\{1,2\}$, *is* $\emptyset \to \{1\} \to \{1,2\}$. *That is, the number of paths from* $\emptyset$ *to* $\{1,2\}$ *is 2, while the number of paths via* $\{1\}$ *is 1. Then player 1 obtains* $\frac{1\ path}{2\ paths}$ *of the amount of the Harsanyi dividend* $\Delta^v(12)$ *(i.e.,* $\frac{1}{2}\Delta^v(12)$*). In the same way, player 1 obtains* $\frac{1}{1}\Delta^v(1)$ *and* $\frac{0}{1}\Delta^v(2)$. *The Shapley value of player 1 is obtained as the sum of all these shares, i.e.,* $\frac{1}{1}\Delta^v(1) + \frac{0}{1}\Delta^v(2) + \frac{1}{2}\Delta^v(12)$. *This allocation rule can be extended to the case* $N = \{1,2,3\}$, *e.g., there are six shortest paths from* $\emptyset$ *to* $\{1,2,3\}$ *(see, Fig. 7* (b)*). Among them, two paths,* $\emptyset \to \{1\} \to \{1,2\} \to \{1,2,3\}$ *and* $\emptyset \to \{1\} \to \{1,3\} \to \{1,2,3\}$, *pass through* $\{1\}$. *Then, the following holds:*

$$\phi_1(\{1,2,3\},v) = \frac{1}{1}\,\Delta^v(1) + \frac{1}{2}\,\Delta^v(12) + \frac{1}{2}\,\Delta^v(13) + \frac{0}{2}\,\Delta^v(23) + \frac{2}{6}\,\Delta^v(123).$$

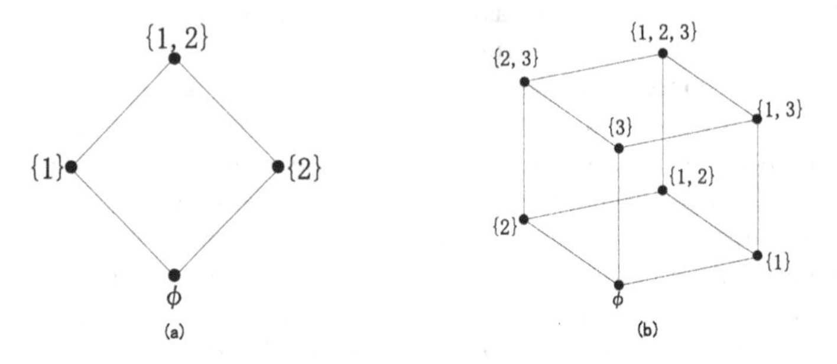

Fig. 7. The Boolean lattice $B(2)$ and $B(3)$

Definition 39 (sharing system [9]). Let (N, v, E) be a communication situation consisting of a game (N, v) and a communication network $G := (N, E)$. A *sharing system* on $\mathfrak{F}(G)$ is a system $p := (p^S)_{S\in\mathfrak{F}(G)}$, where p^S is a s-dimensional vector assigning a non-negative share $p_i^S \geq 0$ to every player $i \in S$ with $\sum_{j\in S} p_j^S = 1$, for any $S \in \mathfrak{F}(G)$.

Definition 40 (the Harsanyi solution [9]). We denote the collection of all sharing systems on $\mathfrak{F}(G)$ by $\mathfrak{S}^G$. For a communication situation $(N, v, E) \in CS^N$ and a sharing system $p \in \mathfrak{S}^G$, the corresponding *Harsanyi payoff vector* is the payoff vector $H^p(N, v, E) \in \mathbb{R}^N$ given by

$$H_i^p(N,v,E) = \sum_{i\in S\in\mathfrak{F}(G)} p_i^S\, \Delta^{\mathfrak{F}(G)}(S) = \sum_{i\in S\in\mathfrak{F}(G)} p_i^S\, \Delta^{v^E}(S) \quad \forall i \in N.$$

A *Harsanyi solution*, as an allocation rule, assigns for a given sharing system $p \in \mathfrak{S}^G$ the Harsanyi payoff vector $H^p(N, v, E)$ to each communication situation (N, v, E). Due to the equality $v(N) = \sum_{S \in \mathfrak{F}(G)} \Delta^{(N,v,E)}(S)$, we have

$$\sum_{i \in N} H_i^p(N, v, E) = v(N),$$

and thus each Harsanyi solution is efficient. The Shapley value is the Harsanyi solution that assigns to any communication situation $(N, v, \{ij \mid i, j \in N, i \neq j\})$ (i.e., to any ordinary cooperative game (N, v)), the Harsanyi payoff vector $H^p(N, v, \{ij \mid i, j \in N, i \neq j\})$ with the sharing system p given by $p_i^S = \frac{1}{s}$ for each $S \in \mathfrak{F}(G)$ containing i.

Definition 41 (Harsanyi power solution [9]). A *power measure* on graph $G = (N, E)$ is a function q that assigns to any subgraph $G(S) = (S, E(S))$, $S \subseteq N$ a non-negative vector $q(S, E(S)) \in \mathbb{R}_+^S$, yielding the non-negative power $q_i(S, E(S))$ of each node $i \in S$ in the graph $G(S)$. Then, given a positive power measure q, we can define the corresponding *Harsanyi power solution*, denoted by $H^{p(q)}(N, v, E)$, through the sharing system $p(q) = (p^S(q))_{S \in \mathfrak{F}(G)}$ induced by the power measure q as

$$p_i^S(q) = \frac{q_i(S, E(S))}{\sum_{j \in S} q_j(S, E(S))}$$

for all $i \in S$ whenever $\sum_{j \in S} q_j(S, E(S)) \neq 0$ and $p_i^S(q) = \frac{1}{s}$ if $\sum_{j \in S} q_j(S, E(S)) = 0$.

A characteristic of the Harsanyi power solutions for communication situations is that we associate a sharing system with some *power measure*, being a function which assigns a non-negative real number to every node in the graph, for the underlying communication networks. These numbers represent the strength or power of those nodes in the graph. Given a power measure we define the corresponding sharing system such that the share vectors for every coalition are proportional to the power measure of the corresponding subgraph.

Social network researchers have considered some fundamental properties of the individuals, that inform us about specific factors such as who is who in the network: who is leader, who is intermediary, who is nearly isolated, who is central, and who is peripheral. Here, we introduce several concepts of *degree* of *centrality* and *peripherality* for a node (position, actor, individual) in a network (undirected graph) as examples of power measures on undirected graphs [16,19,17].

Centrality measures [16,19]. *Centrality* is a sociological concept which is not clearly defined; it is frequently defined only in an undirected manner. For example, the literature presents several alternative definitions for centrality. We review some of these definitions below:

(Dc: degree centrality). It measures the degree to which an actor i can communicate directly with other actors:

$$q_i^{DEG}(S, E(S)) := \sum_{j \in S} a^{G(S)}(i).$$

(Cc: closeness centrality). It measures the degree to which an actor i is close to other actors:

$$q_i^{CLO}(S, E(S)) := \sum_{j \in S \setminus \{i\}} \frac{1}{d_{ij}^{G(S)}}.$$

(Bc: betweenness centrality). It measures the degree to which an actor i lies on the shortest paths between other actors in the network:

$$q_i^{BET}(S, E(S)) := \sum_{j,k \in S \setminus i} \frac{\#\text{ of geodesics in } S \text{ from } j \text{ to } k \text{ via } i}{\#\text{ of geodesics in } S \text{ from } j \text{ to } k}.$$

(Oc: originator centrality). It measures the degree to which an actor i is required as an initiator/originator in network-forming processes:

$$q_i^{ORI}(S, E(S)) := \frac{||i \to S||}{||\emptyset \to S||},$$

where $||i \to S||$ (resp., $||\emptyset \to S||$) means the number of shortest paths from i (resp., $\emptyset$) to S in the Hasse diagram of the poset induced by a graph $(S, E(S))$.

Peripherality Measure [17]. With regard to networks such as roads, railways, airways, the Internet, and others that use nodes or terminals such as airports and railway stations, etc, terminal cities/nodes benefits far more from direct/indirect access to big cites (important nodes or central hubs) than do big cities receive from connecting to terminal cities/nodes. Indeed, peripheral cities bear a heavier burden than central cities in the construction/extension of highways/railways. Fujimoto [17] has proposed a *peripherality measure* on undirected graphs axiomatically as follows.

$$q_i^{PER}(S, E(S)) := \frac{||\emptyset \to S \setminus i||}{||\emptyset \to S||}.$$

Note (The Myerson and Position Values as Harsanyi Power Solutions): Brinks et al. [9] pointed out and demonstrated that the Myerson and position values are typical Harsanyi power solutions with simple power measures for some types of communication situations.

Let (N, v, E) be a communication situation consisting of a game (N, v) and a communication network $G := (N, E)$. The Myerson value $\Psi(N, v, E)$ is the Harsanyi power solution with the sharing system p induced by the *egalitarian power measure* q^E, e.g.,

$$q_i^E(S, E(S)) = 1 \quad \forall S \in \mathfrak{F}(G),\ \forall i \in S,$$

i.e.,

$$p_i^S = \frac{1}{s} \quad \forall S \in \mathfrak{F}(G),\ \forall i \in S.$$

If (N, E) is cycle free, the position value $\pi(N, v, E)$ is the Harsanyi solution with the sharing system p induced by the *degree centrality measure* $q^{DEG}(S, E(S))$.

All the power measures, q^E, q^{DEG}, q^{CLO}, q^{BET}, q^{ORI}, and q^{PER} induce the Shapley value under complete communication situations.

4.4 Numerical Examples

In this subsection, we make comparisons among the existing five types of Harsanyi power solutions (the Shapley value ϕ, the Myerson value Ψ, the position value π, the Harsanyi power solutions induced by the originate centrality measure Φ^{ORI} and the peripherality measure Φ^{PER}) in some communication situations. The Harsanyi dividend of any coalition which is not contained within any connected component of the communication network in a communication situation is always zero, i.e., $\Delta^{(N,v,E)}(S) = 0$ if $S \not\subseteq C$ for any $C \in N/E$. Therefore, in considering the Harsanyi power solutions for a communication situation (N, v, E), we can assume that the communication network (N, E) is connected without loss of generality. Examples 8, 9, and 10 not only show comparisons of them but also illustrate criticisms against the the Myerson value and/or the position value. Two criticisms are reproduced below (see, e.g., [37] for additional details):

On the Myerson value :

$$\Psi_i(N, u_S, E_1) = \Psi_i(N, u_S, E_2) = \frac{1}{|S|} \quad \forall i \in N$$

whenever S is a feasible coalition in both (N, E_1) and (N, E_2), where u_S is the unanimity game of S. Furthermore, in the communication situation with $E^* = \{ij \subseteq N \mid j \in N \setminus i\}$ (i.e., E^* is a star-shape graph with a central player i), every player receives the same value (see $\Psi(N, v, E_e)$ in Example 10).

On the position value :

Irrelevant null players often have positive values (see Example 9). Recall that a null player $i \in N$ of the game (N, v) is a player satisfying that $v(S \cup i) = v(S)$ for any $S \subseteq N \setminus i$.

Example 8. Consider the communication situation (N, v, E) with $N = \{1, 2, 3\}$, $E = \{13, 23\}$ (Fig. 5 (b)), and

$$v(S) = \begin{cases} 0 & \text{if } |S| \leq 1, \\ 30 & \text{if } |S| = 2, \\ 36 & \text{if } S = N. \end{cases}$$

Then,

$$\begin{aligned} \phi(N, v) &= (12, 12, 12), \\ \Psi(N, v, E) &= (7, 7, 22), \\ \pi(N, v, E) &= (9, 9, 18), \\ \Phi^{ORI}(N, v, E) &= (9, 9, 18), \\ \Phi^{PER}(N, v, E) &= (3, 3, 30). \end{aligned}$$

Example 9. Consider the communication situation (N, v, E) with $N = \{1, 2, 3\}$, $E = \{12, 13, 23\}$ (Fig. 5 (a)), and

$$v(S) = \begin{cases} 12 & \text{if } S \supseteq \{1, 2\}, \\ 0 & \text{otherwise.} \end{cases}$$

That is, the player 3 is a null player. Then,

$$\phi(N, v) = (6, 6, 0),$$
$$\Psi(N, v, E) = (6, 6, 0), \quad \pi(N, v, E) = (5, 5, 2),$$
$$\Phi^{ORI}(N, v, E) = (6, 6, 0), \quad \Phi^{PER}(N, v, E) = (6, 6, 0).$$

Example 10. Consider communication situations (N, u_N, E) with connected graphs in Fig.8 which shows all connected graphs (up to isomorphism) with $2 \le n \le 4$ nodes. Then, for any such communication situations,

$$\phi_i(N, u_N, E) = \Psi_i(N, u_N, E) = \frac{1}{n} \quad \forall i \in N.$$

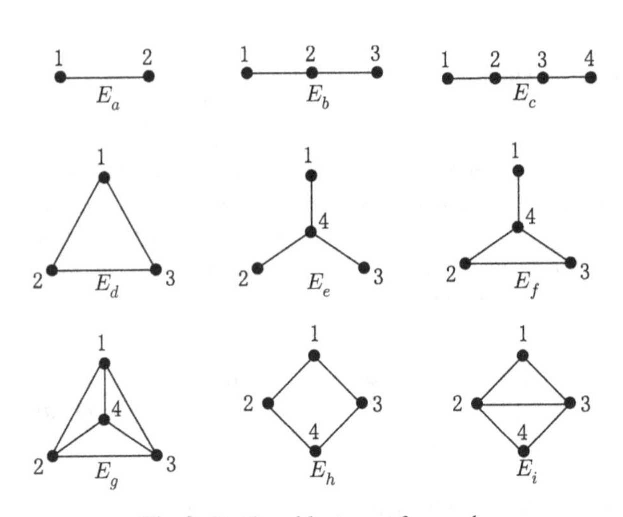

Fig. 8. Graphs with at most four nodes

Table 1 displays the remaining values (i.e., the position value π, the Harsanyi power solutions induced by the originate centrality measure Φ^{ORI} and the peripherality measure Φ^{PER}).

5 Cooperative Game and Combinatorial Structure

In section 4, we considered the following case:

> **Some subsets of N may not be meaningful.** *When N is the set of political parties, it means that some coalitions of parties are unlikely to occur, or even impossible (coalition mixing left and right parties); When N is the set of players, for players in order to coordinate their actions, they must be able to communicate [13,22,61].*

Table 1. Comparison of existing values

	π	Φ^{ORI}	Φ^{PER}
E_a	$(\frac{1}{2}, \frac{1}{2})$	$(\frac{1}{2}, \frac{1}{2})$	$(\frac{1}{2}, \frac{1}{2})$
E_b	$(\frac{1}{4}, \frac{1}{2}, \frac{1}{4})$	$(\frac{1}{4}, \frac{1}{2}, \frac{1}{4})$	$(\frac{1}{2}, 0, \frac{1}{2})$
E_c	$(\frac{1}{6}, \frac{2}{6}, \frac{2}{6}, \frac{1}{6})$	$(\frac{1}{8}, \frac{3}{8}, \frac{3}{8}, \frac{1}{8})$	$(\frac{1}{2}, 0, 0, \frac{1}{2})$
E_d	$(\frac{1}{3}, \frac{1}{3}, \frac{1}{3})$	$(\frac{1}{3}, \frac{1}{3}, \frac{1}{3})$	$(\frac{1}{3}, \frac{1}{3}, \frac{1}{3})$
E_e	$(\frac{1}{6}, \frac{1}{6}, \frac{1}{6}, \frac{3}{6})$	$(\frac{1}{6}, \frac{1}{6}, \frac{1}{6}, \frac{3}{6})$	$(\frac{1}{3}, \frac{1}{3}, \frac{1}{3}, 0$
E_f	$(\frac{3}{12}, \frac{2}{12}, \frac{2}{12}, \frac{5}{12})$	$(\frac{2}{14}, \frac{3}{14}, \frac{3}{14}, \frac{6}{14})$	$(\frac{3}{7}, \frac{2}{7}, \frac{2}{7}, 0)$
E_g	$(\frac{1}{4}, \frac{1}{4}, \frac{1}{4}, \frac{1}{4})$	$(\frac{1}{4}, \frac{1}{4}, \frac{1}{4}, \frac{1}{4})$	$(\frac{1}{4}, \frac{1}{4}, \frac{1}{4}, \frac{1}{4})$
E_h	$(\frac{1}{4}, \frac{1}{4}, \frac{1}{4}, \frac{1}{4})$	$(\frac{1}{4}, \frac{1}{4}, \frac{1}{4}, \frac{1}{4})$	$(\frac{1}{4}, \frac{1}{4}, \frac{1}{4}, \frac{1}{4})$
E_i	$(\frac{13}{60}, \frac{17}{60}, \frac{17}{60}, \frac{13}{60})$	$(\frac{2}{10}, \frac{3}{10}, \frac{3}{10}, \frac{2}{10})$	$(\frac{3}{10}, \frac{2}{10}, \frac{2}{10}, \frac{3}{10})$

In this section, we elaborate on more general cases, including the case discussed in section 3, as follows:

> **Subsets of N may not be "black and white [22] ",** *which means that the membership of an element to N may not be simply a matter of member or non-member. This is the case with multi-criteria decision making when underlying scales are bipolar, i.e., a central value exists on each scale, which is a demarcation between values considered as "good", and as "bad", the central value being neutral; In voting situation, it is convenient to consider that players may also abstain, hence each voter has three possibilities [15]; When N is the set of players, one may consider that each player can play at different level of participation [36].*

5.1 Generalization of Domains of Cooperative Games

Definition 42 (lattice). Let L be a non empty set and $\leq$ a partial order on L (i.e., $(L, \leq)$ is a poset). A poset $(L, \leq)$ is said to be a *lattice* if for $x, y \in L$, the supremum $x \vee y$ and the infimum $x \wedge y$ always exist. $\top$ and $\perp$ are the top (greatest) and bottom (least) elements of L, if they exist. An element $j \in L$ is said to be *join-irreducible* if it is not $\perp$ and cannot be express as a supremum of other elements (i.e., there are no $i, k < j$ such that $j = i \vee k$). The set of all join-irreducible elements of L is denoted by $\mathfrak{J}(L)$. A lattice $(L, \leq)$ is *distributive* if $\vee, \wedge$ obey distributivity. We often identify a lattice $(L, \leq)$ with L or with $(L, \leq, \vee, \wedge, \top, \perp)$.

Definition 43 (cooperative game on lattice). A pair (L, v) consisting of a lattice L and a (characteristic) function $v : L \to \mathbb{R}$ such as $v(\perp) = 0$ constitutes a *cooperative game on a lattice.*

The power set 2^N of N can coincide with the Boolean lattice $B(n)$. Therefore, an ordinary cooperative game (N, v) is regarded as a *cooperative game on a lattice* $((2^N, \subseteq), v)$. Indeed, the infimum (bottom element) in the lattice $(2^N, \subseteq)$ is the empty set $\emptyset$ and $v(\emptyset) = 0$. A communication situation (N, v, E) also can be regarded as a *cooperative game on a lattice*, if the communication network (N, E) is connected, because the poset induced by the connected graph (N, E) is obviously a lattice with $\emptyset$ as the bottom element. However, a bi-cooperative game $(N, v, \mathcal{Q}(N))$ is generally not a *cooperative game on a lattice*. Indeed, the family $\mathcal{Q}(N)$ with $(\mathcal{Q}(N), \sqsubseteq, \sqcup, \sqcap, (N, \emptyset), (\emptyset, N))$ is a lattice with $(\emptyset, N)$ as the bottom element, but $v(\emptyset, N)$ is not always zero.

Proposition 9 **([6]).** *Let L be a distributive lattice. Any element $x \in L$ can be written as an irredundant supremum of join-irreducible elements in a unique way. That is, for any $x \in L$ there uniquely exists $\{j_1, \ldots, j_m\} \subseteq \mathfrak{J}(L)$ such that*

$$x = \bigvee_{i=1}^{m} j_i \tag{11}$$

and that if there exists $M \subseteq \mathfrak{J}(L)$ such that $x = \bigvee_{j \in M} j$, then $M \supseteq \{j_1, \ldots, j_m\}$. The equation (11) is called the minimal decomposition *of x and the $\{j_1, \ldots, j_m\}$ is denoted by $\eta^*(x)$. For any $x \in L$, we denote by $\eta(x) := \{j \in \mathfrak{J}(L) \mid j \leq x\}$, then $x = \bigvee_{j \in \eta(x)} j$. For example, in Fig. 9 (b), $\eta(23, 1) = \{(\emptyset, 13), (2, 13), (\emptyset, 12), (3, 12)\}$ and $\eta^*(23, 1) = \{(2, 13), (3, 12)\}$.*

Theorem 1 **(Birkhoff's theorem [6]).** *For any poset $(P, \leq)$, a subset $Q \subseteq P$ is said to be a* down set *of P if $x \in Q$ and that $y \leq x$ implies $y \in Q$. We denote by $\mathcal{O}(P)$ the set of all downsets of P. One can associate to any poset $(P, \leq)$ a distributive lattice which is $\mathcal{O}(P)$ endowed with inclusion. Then, for any lattice L, the mapping η is an isomorphism of L onto $\mathcal{O}(\mathfrak{J}(L))$.*

5.2 Examples of Generalizations of Games [22]

This subsection shows some examples of *cooperative games on lattices*.

Restricted Domains

Definition 44 **(game on convex geometry [3]).** Let N be a set of players. A collection $\mathcal{CG}$ of subsets of N is called a *convex geometry* if (i) it contains the empty set, (ii) it is closed under intersection, and (iii) $S \in \mathcal{CG}$, $S \neq N$ implies that there exists $j \in N \setminus S$ such that $S \cup j \in \mathcal{CG}$. A *cooperative game on a convex geometry* $\mathcal{CG}$ is a triplet $(N, v, \mathcal{CG})$ with a function $v : \mathcal{CG} \to \mathbb{R}$ such that $v(\emptyset) = 0$. In addition, several other games on restricted domains (e.g., *union stable systems*, *matroids*, and so on), which are generalization of posets induced by connected graphs, also have been proposed and studied by Bilbao [3].

Extended Domains

Definition 45 (multichoice game [36]). Let N be a set of players. Each player $i \in N$ has a finite number of feasible participation levels whose set we denote by $M_i = \{0, 1, \ldots, m_i\}$ and $\mathcal{M} = \prod_{i \in N} M_i$. Each element $\mathbf{s} = (s_1, s_2, \ldots, s_n) \in \mathcal{M}$ specifies a *participation profile* for players and is referred to as a *multichoice coalition*. So, a multichoice coalition indicates the participation level of each player. A triplet $(N, v, \mathcal{M})$ consisting of a (characteristic) function $v : \mathcal{M} \to \mathbb{R}$ such that $v(\mathbf{0}) = 0$, where $\mathbf{0} = (0, 0, \ldots, 0) \in \mathcal{M}$, constitutes a *multichoice game*.

Definition 46 (game on direct product of distributive lattices [46]). Let $N = \{1, \cdots, n\}$ be a finite set, $\{L_i\}_{i \in N}$ a set of distributive lattices and $\boldsymbol{L} := \prod_{i \in N} L_i$. (Notice that $\boldsymbol{L}$ is also a distributive lattice with the product order induced by $\{L_i\}_{i \in N}$). A triplet $(N, v, \boldsymbol{L})$ consisting of a product lattice $\boldsymbol{L} = \prod_{i \in N} L_i$ and a (characteristic) function $v : \boldsymbol{L} \to \mathbb{R}$ such as $v(\bot_1, \cdots, \bot_n) = 0$, where $\bot_i$ is the bottom element of L_i for each $i \in N$, constitutes a *cooperative game on a direct product of distributive lattices*.

Here, we consider some examples of games on a direct product of distributive lattices (see, also Fig. 9). If L_i is a two-element lattice (i.e., $L_i := \{\bot_i, \top_i\}$) for all players $i \in N$, then we get ordinary games on 2^N (Fig. 9 (a)); If $L_i := \{0, 1, \ldots, m_i\}$ for all players $i \in N$, we obtain multichoice games on $\mathcal{M} = \prod_{i \in N} L_i$ (Fig. 9 (c)); If $L_i := \{\bot_i, x_i, \top_i\}$, $\bot_i < x_i < \top_i$ (e.g., $\{-1, 0, 1\}$) for all players $i \in N$, then the product lattice $\boldsymbol{L} := \prod_{i \in N} L_i$ is isomorphic to $(\mathcal{Q}(N), \sqsubseteq)$ (Fig. 9 (b)). The bottom element $(\bot_1, \cdots, \bot_n)$ in $\boldsymbol{L}$ corresponds to $(\emptyset, N)$ in $\mathcal{Q}(N)$. That is, a bi-cooperative game $(N, v, \mathcal{Q}(N))$ is generally not regarded as a game on a direct product of distributive lattices since bi-cooperative games need not be vanishing at the bottom element $(\emptyset, N)$.

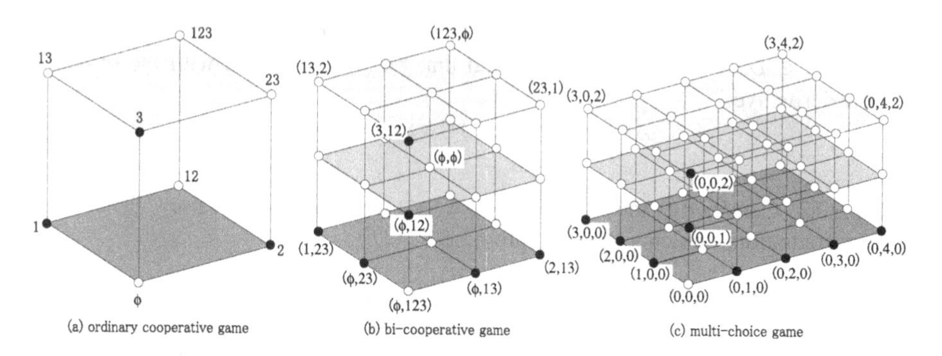

Fig. 9. Examples of direct products of distributive lattices: elements indicated by black circles are join-irreducible

Generally, the Möbius transform Δ^v of a game v on a lattice L can be implicitly defined through Definition 3. As it will be seen in the next section, *derivatives* of games on lattices are a very useful tool, and have been generalized (in particular) for games on *distributive lattices* (see, [26]).

Definition 47 (discrete derivative on distributive lattice). Let L be a distributive lattice. The *first order derivative* of a game $v : L \to \mathbb{R}$ with respect to a join-irreducible element $i \in \mathfrak{J}(L)$ at $x \in L$ is given by

$$\Delta_i v(x) := v(x \vee i) - v(x).$$

The *derivative* of v with respect to $y \in L$ at $x \in L$ is iteratively defined by

$$\Delta_y v(x) := \Delta_{j_1}[\Delta_{j_2}[\cdots \Delta_{j_{m-1}}[\Delta_{j_m} v(x)]\cdots]],$$

where $\eta^*(y) = \{j_1, j_2, \ldots, j_m\} \subseteq \mathfrak{J}(L)$ is the minimal decomposition of y. Note that if $j_k \leq x$ for some k, the derivative is null. Also, $\Delta_y v(x)$ does not depend on the order of the j_k's. The explicit formula is:

$$\Delta_y v(x) = \sum_{S \subseteq \{1,\ldots,m\}} (-1)^{m-s}\, v(x \vee \bigvee_{k \in S} j_k), \tag{12}$$

equivalently,

$$\Delta_y v(x) = \sum_{y \leq z \leq x \vee y} \Delta^v(z).$$

In particular,

$$\Delta_y v(\bot) = \Delta^v(y) \quad \forall y \in L.$$

Example 11. An ordinary game (N, v) can be regarded as a cooperative game on the lattice $L = (2^N, \subseteq)$. The join operator in the lattice is $\cup$ operator. The set of all join-irreducible elements $\mathfrak{J}(L)$ is N, i.e., any $i \in N$ is a join-irreducible element. For any $T \subseteq N$ (i.e., $T \in L$), the minimal decomposition of T is T itself, i.e., $\eta^*(T) = T = \{\sigma_T(1), \ldots, \sigma_T(t)\}$ for some permutation σ_T on N. An order $\{\sigma_T(k)\} \leq U$ coincides with $\sigma_T(k) \in U$. Then, we can easily find that Eq.(12) coincides with the ordinary discrete derivative:

$$\Delta_T v(U) = \sum_{S \subseteq T} (-1)^{|T|-|S|}\, v(U \cup S).$$

Example 12. Considering the lattice $L = (\mathcal{Q}(N), \sqsubseteq)$ with the join operator $\sqcup$. The set of all join-irreducible elements $\mathfrak{J}(L)$ is represented by

$$\{\{(i, N \setminus i)\}_{i \in N}, \{(\emptyset, N \setminus i)\}_{i \in N}\}.$$

For any $(T_1, T_2) \in \mathcal{Q}(N)$ (i.e., $(T_1, T_2) \in L$), the minimal decomposition of (T_1, T_2) is represented by

$$\eta^*(T_1, T_2) = \left\{\{(i, N \setminus i)\}_{i \in T_1}, \{(\emptyset, N \setminus j)\}_{j \in N \setminus (T_1 \cup T_2)}\right\}.$$

However, the discrete derivative in a bi-cooperative game $(N, v, \mathcal{Q}(N))$ (see, Eq. (7) in Definition 24) does not coincide with that in the cooperative game on the lattice $(\mathcal{Q}(N), \sqsubseteq)$. Indeed, for $N = \{1, 2, 3\}$, the $(123, 0)$-derivative at $(\emptyset, 3)$ can be defined in a cooperative game on the lattice $(\mathcal{Q}(\{1, 2, 3\}), \sqsubseteq)$, but cannot in the bi-cooperative game $(\{1, 2, 3\}, v, \mathcal{Q}(\{1, 2, 3\}))$. Now, let $\Delta^B_{(T_1,T_2)}$ denote the (T_1, T_2)-derivative in the sense of bi-cooperative games and $\Delta^L_{(T_1,T_2)}$ the (T_1, T_2)-derivative in the sense of cooperative games on lattices. For $(A_1, A_2), (B_1, B_2) \in Q(N)$ such as $(A_1, A_2) \sqsubseteq (B_1, B_2)$ and $A_2 \cap B_1 = \emptyset$, we consider the following formula:

$$\Delta^v([(A_1, A_2), (B_1, B_2)]) := \sum_{(A_1,A_2)\sqsubseteq(S_1,S_2)\sqsubseteq(B_1,B_2)} (-1)^{|S_1\setminus A_1|+|S_2\setminus B_2|}\, v(S_1, S_2). \tag{13}$$

Then, for $(T_1, T_2) \in Q(N)$ and $(S_1, S_2) \in Q(N \setminus (T_1 \cup T_2))$,

$$\Delta^B_{(T_1,T_2)} v(S_1, S_2 \cup T_2) = \Delta^v([(S_1, S_2 \cup T_2), (S_1 \cup T_1, S_2)]).$$

For $(A_1, A_2), (B_1, B_2) \in Q(N)$ such as $(A_1, A_2) \sqsubseteq (B_1, B_2)$ and $A_2 \cap B_1 = \emptyset$,

$$\Delta^v([(A_1, A_2), (B_1, B_2)]) = \Delta^B_{(B_1\setminus A_1, A_2\setminus B_2)} v(A_1, A_2) = \Delta^L_y v(A_1, A_2),$$

where $y = \bigsqcup_{(S_1,S_2)\in\eta^*(B_1,B_2)\setminus\eta^*(A_1,A_2)} (S_1, S_2)$. For example (see, Fig. 2),

$$\begin{aligned}\Delta^v([(\emptyset, 3), (1, \emptyset)]) &= v(1, \emptyset) - v(1, 3) - v(\emptyset, \emptyset) + v(\emptyset, 3)\\ &= \Delta^B_{(1,3)} v(\emptyset, 3)\\ &= \Delta^L_{(1,2)} v(\emptyset, 3).\end{aligned}$$

5.3 Value and Interaction Index in Games on Distributive Lattices

In this subsection, we discuss on a specific type of game on lattice, where the lattice is a direct product of distributive lattices. Let $N := \{1, \ldots, n\}$ and $\boldsymbol{L} := L_1 \times \cdots \times L_n$, where $L_1, \ldots, L_n$ are finite distributive lattices. Then, $\boldsymbol{L}$ is also a distributive lattice and all join-irreducible elements of $\boldsymbol{L}$ are of the form $(\bot_1, \ldots, \bot_{i-1}, j_i, \bot_{i+1}, \ldots, \bot_n)$ for some $i \in N$ and some $j_i \in \mathfrak{J}(L_i)$. A *vertex* of $\boldsymbol{L}$ is any element whose components are either top or bottom. Vertices of $\boldsymbol{L}$ will be denoted by $\top_Y$, $Y \subseteq N$, whose coordinates are $\top_k$ if $k \in Y$, $\bot_k$ otherwise, for $k \in N$. Each lattice L_i represents the poset of action, choice, or participation level of player $i \in N$ to the game. An ordinary cooperative game (N, v) can be regarded as the following game $v^{\boldsymbol{L}} : \boldsymbol{L} \to \mathbb{R}$:

> Let $L_i := \{0, 1\}$ with the ordinary order $\leq$ on integers for all $i \in N$, and $\boldsymbol{L} = \prod_{i\in N} L_i$. Then, $\mathfrak{J}(L_i) = \{1\}$ for all $i \in N$. So, $\bot_i = 0$ and $\top_i = 1$ for all $i \in N$, therefore $\top_S = \chi_S$. Moreover, for any $y \in \boldsymbol{L}$ there uniquely exists $Y \subseteq N$ such that $y = \chi_Y = \top_Y$. Thus,
>
> $$v^{\boldsymbol{L}}(\top^Y) := v(Y) \quad Y \subseteq N$$
>
> is the desired one.

Lange and Grabisch [46] give the following interpretation for games on $\boldsymbol{L}$:

> *We assume that each player i $\in$ N has at her/his disposal a set of* elementary or pure actions $j_1, \cdots, j_{j_i}$. *These elementary actions are partially ordered (e.g., in the sense of benefit caused by the action), forming a partially ordered set*

$(\mathfrak{E}_i, \leq_i)$, $\mathfrak{E}_i = \{j_1, \cdots, j_{j_i}\}$. Then by Birkhoff's theorem (Theorem 9), the set $(\mathcal{O}(\mathfrak{E}_i), \subseteq)$ of downsets of $\mathfrak{E}_i$ is a distributive lattice denoted by L_i, whose join-irreducible elements correspond to the elementary actions. The bottom action $\bot_i$ of L_i is the action which amounts to do nothing. Hence, each action in L_i is either a pure action j_k or a combined action $j_k \vee j_{k'} \vee j_{k''} \vee \cdots$ consisting of doing all pure actions $j_k, j_{k'}, j_{k''} \cdots$ for player $i \in N$.

For a given elementary action $j_i \in \mathfrak{J}(L_i) \subseteq L_i$, the importance index (the (Shapley-type) value) of a game v on a direct product lattice $\boldsymbol{L} = \prod_{i \in N} L_i$ of distributive lattices $\{L_i\}_{i \in N}$ is written as a weighted average of the marginal contributions of j_i, taken at vertices of $\boldsymbol{L}$. This important index has been a generalization of the Shapley value in both ordinary games and multichoice games.

Definition 48 (importance index). *Let $i \in N$ and $j_i \in \mathfrak{J}(L_i)$. The* importance index *with respect to j_i of a game $v : \boldsymbol{L} \to \mathbb{R}$ is defined by*

$$\phi_{j_i}(v) := \sum_{Y \subseteq N \setminus i} \frac{y!(n-y-1)!}{n!} \Delta_{j_i} v(\top_Y).$$

As an extension of the importance index for every element of $\boldsymbol{L}$ and every game $(N, v, \boldsymbol{L})$, the interaction transform on $\boldsymbol{L}$ has been proposed by Lange and Grabisch [44]. For any $\boldsymbol{x} \in \boldsymbol{L}$, $I_{\boldsymbol{x}}(N, v, \boldsymbol{L})$ expresses the interaction in the game among all elementary actions $\boldsymbol{j}$ of the minimal decomposition $\boldsymbol{x} = \bigvee_{\boldsymbol{j} \in \eta^*(\boldsymbol{x})} \boldsymbol{j}$.

Definition 49 (antecessors). The *antecessor* $\underline{\boldsymbol{x}}$ of $\boldsymbol{x} \in \boldsymbol{L}$ is defined as

$$\underline{\boldsymbol{x}} = \bigvee \{j \in \eta(\boldsymbol{x}) \mid j \notin \eta^*(\boldsymbol{x})\}$$

with convention $\underline{\bot} = \bot$ and $\bigvee \emptyset = \bot$. If $\boldsymbol{x}$ is a join-irreducible element (i.e., $\boldsymbol{x} \in \mathfrak{J}(\boldsymbol{L})$), the antecessor of $\boldsymbol{x}$ is obviously its predecessor, in accordance with the notation $\underline{\boldsymbol{x}}$. Note also that the definition $\underline{\boldsymbol{x}}$ is consistent with the structure of each lattices L_i. Indeed, $\underline{\boldsymbol{x}} = (\underline{x_1}, \cdots, \underline{x_n})$.

Definition 50 (interaction transform on product lattices [44]). The *(Shapley-type) interaction transform* $I_{\boldsymbol{x}}(N, v, \boldsymbol{L})$ with respect to $\boldsymbol{x} \in \boldsymbol{L}$ of $v : \boldsymbol{L} \to \mathbb{R}$ is defined by

$$I_{\boldsymbol{x}}(N, v, \boldsymbol{L}) := \sum_{Y \subseteq N \setminus X} \frac{|Y|!\,(n-|X|-|Y|)!}{(n-|X|+1)!} \Delta_{\boldsymbol{x}} v(\underline{\boldsymbol{x}} \vee \top_Y),$$

where $X = \{i \in N \mid x_i \neq \bot_i\}$. Equivalently,

$$I_{\boldsymbol{x}}(N, v, \boldsymbol{L}) = \sum_{\boldsymbol{x} \leq \boldsymbol{z} \leq \boldsymbol{x}^{\bot}} \frac{1}{k(\boldsymbol{z}) - k(\boldsymbol{x}) + 1} \Delta^v(\boldsymbol{z}),$$

where $\boldsymbol{x}^{\bot} := \top_i$ if $x_i = \bot_i$ and $x^{\bot} := x_i$ if $x_i \neq \bot_i$, and $k(\boldsymbol{y}) = |\{i \in N \mid y_i \neq \bot_i\}|$. Recall that any direct product $\boldsymbol{L} = \prod_{i \in N} L_i$ of distributive lattices $\{L_i\}_{i \in N}$ also a distributive lattice. Thus, the Möbius transform $\Delta^v(\boldsymbol{z})$ and the marginal interaction $\Delta_{\boldsymbol{x}}(\boldsymbol{y})$ in a game $(N, v, \boldsymbol{L})$ can be defined via Definition 47.

Each interaction index in ordinary games, and multichoice games is obtained as a special case of this interaction transform.

5.4 An Importance Index of Games on Regular Set Systems

In this subsection, we introduce an important index on a more general combinatorial structure, which is called the *regular set system* proposed by Honda and Grabisch [35] (see, also [47]). The concept of regular set system is induced by the following condition:

> A condition "*if $S \subsetneq N$ is feasible, then it is possible to find a player $i \in N \setminus S$ such that $S \cup i$ is still feasible*" is one of the weakest restrictions on feasible coalitions in a context where the grand coalition N can form. Because, it says that from a given coalition, it is possible to augment it gradually to reach the grand coalition.

Definition 51 (regular set system). Let us consider $\mathfrak{R}$ a set of coalitions, i.e., $\mathfrak{R} \subseteq 2^N$. Then, a pair $(N, \mathfrak{R})$ is said to be a *set system* on N if $\mathfrak{R}$ contains $\emptyset$ and N, i.e. $\emptyset, N \in \mathfrak{R}$. Elements of $\mathfrak{R}$ are called *feasible coalitions*. For any two feasible coalitions $A \subsetneq B$, we say that *A is covered by* B, and write $A \prec B$, if there is no $C \in \mathfrak{R}$ such that $A \subsetneq C \subsetneq B$. A set system $(N, \mathfrak{R})$ is said to be *regular* if $|B \setminus A| = 1$ whenever $A, B \in \mathfrak{R}$ and $A \prec B$.

Definition 52 (game on regular set system). A triplet $(N, v, \mathfrak{R})$ consisting of a regular set system $(N, \mathfrak{R})$ and a (characteristic) function $v : \mathfrak{R} \to \mathbb{R}$ such as $v(\emptyset) = 0$ constitutes a *game on a regular set system.*

Honda and Fujimoto [34] have proposed axiomatically an importance index of a game on a *regular set system* as a generalization of importance indices of all ordinary games, games on convex geometries, and multichoice games.

Definition 53 (maximal chain of regular set system). Let $\mathfrak{R} \subseteq 2^N$ be a regular set system. If a sequence $C = (C_0, \ldots, C_n)$ satisfies that $C_i \in \mathfrak{R}$ for any $i \in \{0, \cdots, n\}$ and $\emptyset = C_0 \prec C_1 \prec \cdots \prec C_n = N$, then C is called a *maximal chain* of $\mathfrak{R}$. The set of all maximal chains of $\mathfrak{R}$ is denoted by $\mathsf{M}(\mathfrak{R})$.

For any maximal chain $C = (C_0, \ldots, C_n)$, there exists a permutation σ_C on N such that

$$C_i = \bigcup_{k \le i} \{\sigma_C(k)\} \quad \forall i \in \{1, \ldots, n\}. \tag{14}$$

Definition 54 (importance index on regular set system). A *marginal contribution* $\delta_i^v(C)$ of $i \in N$ for a maximal chain $C \in \mathsf{M}(\mathfrak{R})$ in a game $(N, v, \mathfrak{R})$ is defined by

$$\delta_i^v(C) := v(\bigcup_{k \le i} \{\sigma_C(k)\}) - v(\bigcup_{k < i} \{\sigma_C(k)\})$$

where σ_C is a permutation on N satisfying Eq. (14). The *importance index* $\phi(N, v, \mathfrak{R}) \in \mathbb{R}^N$ with respect to a player $i \in N$ of a game $v : \mathfrak{R} \to \mathbb{R}$ on a regular set system $\mathfrak{R}$ is defined by

$$\phi_i(N, v, \mathfrak{R}) := \frac{1}{|\mathsf{M}(\mathfrak{R})|} \sum_{C \in \mathsf{M}(\mathfrak{R})} \delta_i^v(C)$$

for every $i \in N$, where $\phi_i(N, v, \mathfrak{R})$ is the i-th component of $\phi(N, v, \mathfrak{R}) \in \mathbb{R}^N$.

In a case that a regular set system is the power set of N, i.e., $\mathfrak{N} = 2^N$, any game $(N, v, \mathfrak{N})$ coincides with the ordinary game (N, v). Then, $\phi(N, v, \mathfrak{N})$ also coincides with the ordinary Shapley value $\phi(N, v)$. Moreover, through lattice-isomorphic mappings and Birkhoff's Theorem (Theorem 1), this importance index can be applied to games on distributive lattices as the following way :

Definition 55 (set systems induced by lattices). Let $(L, \leq, \vee, \wedge, \top, \bot)$ be a distributive lattice. Then $(L, \leq, \vee, \wedge, \top, \bot) \cong (\eta(L), \subseteq, \cup, \cap, \mathfrak{J}(L), \emptyset)$ with the lattice isomorphism η, where $\eta(x) = \{y \in \mathfrak{J}(L) \mid y \leq x\}$ for $x \in L$, i.e., $\eta(L) = \bigcup_{x \in L}\{\eta(x)\}$ (see, e.g., [6]). Then $(\mathfrak{J}(L), \eta(L))$ is called the *set system* induced by $(L, \leq)$.

All games discussed in this chapter, except bi-cooperative games, can be regarded as games on lattices. All the set systems induced by these lattices become regular [34]. Notice that the set system induced by $(Q(N), \sqsubseteq)$ is also regular. Therefore, we have another representation of importance indices of these games via η as follows:

$$I_{j_i}(N, v, L) := \phi_{\eta(j_i)}\left(\mathfrak{J}(L), v\eta^{-1}, \eta(L)\right) \quad \forall j_i \in \mathfrak{J}(L).$$

6 Concluding Remarks

This chapter shows cooperative games on various extended or restricted domains. We discussed only the Shapley-type values and interaction indices. However, there are various allocation rules and solution concepts in ordinary cooperative game theory, e.g., *the core, bargaining set, prekernel, kernel, prenucleolus, nucleolus, etc.* These various allocation rules and solution concepts can be seen in the literature [11,54]. The *Core* of cooperative games on various domains also have been studied by several researchers (see, e.g., [4,30]). More information about "cooperative game in combinatorial structures" and "social and economic networks in cooperative games" can be found in the literatures (see, e.g., [3,61]). To our knowledge, the topics "interaction indices of games with networks, and on regular set systems" have not been studied yet.

References

1. Aubin, J.P.: Optima and equilibria: An Introduction to Nonlinear Analysis. Springer (1998)
2. Banzhaf, J.F.: Weighted voting doesn't work: A mathematical analysis. Rutgers Law Review 19, 317–343 (1965)
3. Bilbao, J.M.: Cooperative games on combinatorial structures. Kluwer Acadmic Publ., Boston (2000)
4. Bilbao, J.M., Fernández, J.R., Jiménez, N., López, J.J.: The core and the Weber set for bicooperative games. International Journal of Game Theory 36(2), 209–222 (2007)
5. Bilbao, J.M., Fernandez, J.R., Losada, A.J., Lebron, E.: Bicooperative games. In: GAMES 2000, Bilbao, Spain (July 2000)
6. Birkhoff, G.: Lattice Theory, 3rd edn. American Mathematical Society (1967)
7. Borm, P., Owen, G., Tijs, S.: On the position value for communication situations. SIAM Journal on Discrete Mathematics 3, 305–320 (1992)
8. van den Brink, R., Dietz, C.: Multi-player agents in cooperative TU-games. Tinbergen Institute Discussion Paper TI 2012-001/1 (2011)

9. van den Brink, R., van der Laan, G., Pruzhansky, V.: Harsanyi power solutions for graph-restricted games. International Journal of Game Theory 40, 87–110 (2011)
10. Chateauneuf, A., Jaffray, J.Y.: Some characterizations of lower probabilities and other monotone capacities through the use of Möbius inversion. Mathematical Social Sciences 17(3), 263–283 (1989)
11. Curiel, I.: Cooperative game theory and applications. Kluwer Acadmic Publ., Boston (1997)
12. Derks, J., Haller, H., Peters, H.: The selectope for cooperative TU-games. International Journal of Game Theory 29, 23–38 (2000)
13. Dubey, P., Neyman, A., Weber, R.J.: Value theory without efficiency. Mathematics of Operations Research 6, 122–128 (1981)
14. Davey, B.A., Priestley, H.A.: Introduction to Lattices and orders. Cambridge University Press (1990)
15. Felsenthal, D., Machover, M.: Ternary voting games. International Journal of Game Theory 26, 335–351 (1997)
16. Freeman, L.C.: Centrality in network: conceptual classification. Social Network 1, 215–239 (1979)
17. Fujimoto, K.: Network-formation and its stability based on peripherality of nodes in social networks. Journal of Japan Society for Fuzzy Theory and Intelligent Informatics 24(4), 901–908 (2012) (in Japanese)
18. Fujimoto, K., Kojadinovic, I., Marichal, J.L.: Axiomatic characterizations of probabilistic and cardinal-probabilistic interaction indices. Games and Economic Behavior 55, 72–99 (2006)
19. Fujimoto, K., Honda, A.: A value via posets induced by graph-restricted communication situation. In: Proc. 2009 IFSA World Congress/2009 EUSFLAT Conference, Lisbon, Portugal, pp. 636–641 (2009)
20. Fujimoto, K., Murofushi, T.: Some Characterizations of k-Monotonicity Through the Bipolar Möbius Transform in Bi-Capacities. Journal of Advanced Computational Intelligence and Intelligent Informatics 9(5), 484–495 (2005)
21. Grabisch, M.: Modeling data by the Choquet integral. Information Fusion in Data Mining, pp. 135–148. Physica-Verlag, Heidelberg (2003)
22. Grabisch, M.: Capacities and Games on Lattices: A Survey of Results. International Journal of Uncertainty, Fuzziness, and Knowledge-Based Systems 14(4), 371–392 (2006)
23. Grabisch, M., Labreuche, C.: Bi-capacities for decision making on bipolar scales. In: EUROFUSE Workshop on Informations Systems, Varenna, Italy (September 2002)
24. Grabisch, M., Labreuche, C.: Bi-capacities — I: definition, Möbius transform and interaction. Fuzzy Sets and Systems 151, 211–236 (2005)
25. Grabisch, M., Labreuche, C.: Bi-capacities — II: the Choquet integral. Fuzzy Sets and Systems 151, 237–259 (2005)
26. Grabisch, M., Labreuche, C.: Derivative of functions over lattices as a basis for the notion of interaction between attributes. Annals of Mathematics and Artificial Intelligence 49, 151–170 (2007)
27. Grabisch, M., Roubens, M.: An axiomatic approach to the concept of interaction among players in cooperative games. International Journal of Game Theory 28(4), 547–565 (1999)
28. Grabisch, M., Marichal, J.-L., Roubens, M.: Equivalent representations of set functions. Mathematics of Operations Research 25(2), 157–178 (2000)
29. Grabisch, M., Roubens, M.: Application of the Choquet integral in multicriteria decision making. In: Fuzzy Measures and Integrals, pp. 348–374. Physica Verlag, Heidelberg (2000)
30. Grabisch, M., Xie, L.: The core of games on distributive lattices: how to share benefits in a hierarchy. Document de Travail du Centre d'Economie de la Sorbonne (2008.77) (2008)
31. Hamiache, G.: A value with incomplete information. Games and Economic Behavior 26, 59–78 (1999)

32. Harsanyi, J.C.: A bargaining model for the cooperative n-person game. In: Contributions to the Theory of Games IV, pp. 325–355. Princeton University Press, Princeton (1959)
33. Harsanyi, J.C.: A simplified bargaining model for the n-person cooperative game. International Economic Review 4, 59–78 (1963)
34. Honda, A., Fujimoto, K.: A generalization of cooperative games and its solution concept. Journal of Japan Society for Fuzzy Theory and Intelligent Informatics 21(4), 491–499 (2009) (in Japanese)
35. Honda, A., Grabish, M.: Entropy of capacities on lattices. Information Sciences 176, 3472–3489 (2006)
36. Hsiao, C.R., Raghavan, T.E.S.: Shapley value for multi-choice cooperative game. Games and Economic Behavior 5, 240–256 (1993)
37. Jackson, O.M.: Allocation rules for network games. Games and Economic Behavior 51, 128–154 (2005)
38. Kalai, E., Samet, D.: Weighted Shapley values. In: The Shapley Value. Essays in Honor of Lloyd S. Shapley, pp. 83–99. Cambridge Univ. Press (1988)
39. Kojadinovic, I.: Modeling interaction phenomena using non additive measures: Applications in data analysis. PhD thesis, Université de La Réunion, France (2002)
40. Kojadinovic, I.: Estimation of the weights of interacting criteria from the set of profiles by means of information-theoretic functionals. Europ. J. Operational Res. 155, 741–751 (2004)
41. Kojadinovic, I.: A weigh-based approach to the measurement of the interaction among criteria in the framework of aggregation by the bipolar Choquet integral. European Journal of Operational Research 179, 498–517 (2007)
42. Labreuche, C., Grabisch, M.: Modeling positive and negative pieces of evidence in uncertainty. In: Nielsen, T.D., Zhang, N.L. (eds.) ECSQARU 2003. LNCS (LNAI), vol. 2711, pp. 279–290. Springer, Heidelberg (2003)
43. Labreuche, C., Grabisch, M.: Axiomatisation of the Shapley value and power index for bi-cooperative games. Cahiers de la Maison des Sciences Économiques (2006.23) (2006)
44. Lange, F., Grabisch, M.: Games on distributive lattices and the Shapley interaction transform. In: Proc. of IPMU 2008, Torremplinos (Málaga), Spain, pp. 1462–1469 (2008)
45. Lange, F., Grabisch, M.: New axiomatizations of the Shapley interaction index for bi-capacities. Fuzzy Sets and Systems 176, 64–75 (2011)
46. Lange, F., Grabisch, M.: The interaction transform for functions on lattices. Discrete Mathematics 309, 4037–4048 (2009)
47. Lange, F., Grabisch, M.: Values on regular games under Kirchhoff's laws. Mathematical Social Sciences 58, 322–340 (2009)
48. Marichal, J.-L.: Aggregation operators for multicriteria decision aid, Ph.D. thesis, University of Liège (1998)
49. Marichal, J.-L., Roubens, M.: The chaining interaction index among players in cooperative games. In: Meskens, N., Roubens, M. (eds.) Advances in Decision Analysis. Kluwer Acad. Publ., Dordrecht (1999)
50. Murofushi, T., Soneda, S.: Techniques for reading fuzzy measures (iii): interaction index. In: 9th Fuzzy System Symposium, Sapporo, Japan, pp. 693–696 (1993) (in Japanese)
51. Murofushi, T., Sugeno, M.: Fuzzy measures and fuzzy integrals. In: Grabisch, et al. (eds.) Fuzzy Measures and Integrals. Physica-Verlag, Heidelberg (2000)
52. Myerson, R.: Graphs and cooperation in games. Mathematics of Operations Research 2, 225–229 (1977)
53. Owen, G.: Multilinear extensions of games. Management Science 18, 64–79 (1972)
54. Peleg, B., Shudhölter, P.: Introduction to the theory of cooperative games, 2nd edn. Springer, Heidelberg (2007)
55. Ramón, J., Mateo, S.C.: Multi Criteria Analysis in the Renewable Energy Industry. Springer, London (2012)

56. Rota, G.C.: On the foundations of combinatorial theory I. Theory of Möbius functions. Z. Wahrscheinlichkeitstheorie und Verw. Gebiete 2, 340–368 (1964)
57. Roubens, M.: Interaction between criteria and definition of weights in MCDA problems. In: Proc. 44th Meeting of the European Working Group "Multiple Criteria Decision Aiding" (1996)
58. Shafer, G.: A Mathematical Theory of Evidence. Princeton University Press (1976)
59. Shapley, L.: A value for *n*-person games. In: Tucker, A.W., Kuhn, H. (eds.) Contribution to the Theory of Games, II, Annals of Mathematics Studies, vol. 28, pp. 307–317. Princeton Univ. Press, Princeton (1953)
60. Shapley, L.S., Shubik, M.: A Method for Evaluating the Distribution of Power in a Committee System. American Political Science Review 48, 787–792 (1954)
61. Slikker, M., van den Nouweland, A.: Social and economic networks in cooperative game theory. Kluwer Academic Publ., Boston (2001)
62. Sugeno, M.: Theory of fuzzy integrals and its applications. Doctoral Thesis, Tokyo Institute of Technology (1974)
63. Vasil'ev, V.A.: On a class of operators in a space of regular set functions. Optimizacija 28, 102–111 (1982) (in Russian)
64. Vasil'ev, V.A.: Extreme points of the Weber polytope. Discretnyi Analiz i Issledonaviye Operatsyi Ser.1 10, 17–55 (2003) (in Russian)
65. Weber, R.J.: Probabilistic values for games. In: Roth, A.E. (ed.) The Shapley value. Essays in honor of Lloyd S. Shapley, pp. 101–120. Cambridge University Press, Cambridge (1988)

Belief Functions on MV-Algebras of Fuzzy Sets: An Overview

Tommaso Flaminio[1], Lluís Godo[2], and Tomáš Kroupa[3]

[1] Department of Theoretical and Applied Science (DiSTA)
University of Insubria, Via Mazzini 5 - 21100 Varese, Italy
tommaso.flaminio@uninsubria.it

[2] IIIA Artificial Intelligence Research Institute (CSIC)
Campus UAB s/n, Bellaterra 08193, Spain
godo@iiia.csic.es

[3] Institute of Information Theory and Automation of the ASCR
Pod Vodárenskou věží 4, 182 08 Prague, Czech Republic
kroupa@utia.cas.cz

Abstract. Belief functions are the measure theoretical objects Dempster-Shafer evidence theory is based on. They are in fact totally monotone capacities, and can be regarded as a special class of measures of uncertainty used to model an agent's degrees of belief in the occurrence of a set of events by taking into account different bodies of evidence that support those beliefs. In this chapter we present two main approaches to extending belief functions on Boolean algebras of events to MV-algebras of events, modelled as fuzzy sets, and we discuss several properties of these generalized measures. In particular we deal with the normalization and soft-normalization problems, and on a generalization of Dempster's rule of combination.

1 Introduction and Motivations

Dempster-Shafer theory of evidence [13,50] is a generalization of Bayesian probability theory in which degrees of uncertainty are evaluated by belief functions, rather than by probability measures. Belief functions [50,52] can be regarded as a special class of measures of uncertainty used to represent an agent's degrees of confidence in the occurrence of events of interest by taking into account different bodies of evidence that support these beliefs [50]. Such evidence plays a pivotal role in determining the agent's beliefs. Indeed, as we will recall in a while, although any belief function on the Boolean algebra 2^X of subsets of a finite set X might be seen as a particular probability, its associated distribution (called *mass* in Dempster-Shafer theory) maps the whole algebra 2^X into $[0, 1]$, and not only its atoms. Every set $Y \subseteq X$ with a strictly positive mass represents a particular body of evidence and is called a *focal element.*

Given the relevance that Dempster-Shafer theory has in real-world situations, we may argue that usually, when a person is asked to provide an evidence about a fact she were witness of, her description of the facts would be affected not only

V. Torra, Y. Narukawa, and M. Sugeno (eds.), *Non-Additive Measures*,
Studies in Fuzziness and Soft Computing 310,
DOI: 10.1007/978-3-319-03155-2_7,

by uncertainty regarding the statements, but also by a possible *imprecision* in the statements themselves. Therefore the classical framework would be insufficient to model the analysis provided by a witness.

Fuzzy sets were introduced by Zadeh [57] as an extension of classical sets: given a referential set X, a fuzzy subset of X can be identified with a function f from X into the real unit interval $[0, 1]$. Given a fuzzy sets f, the idea is to interpret, for every $x \in X$, the value $f(x) \in [0, 1]$ as the *degree of membership* of x to f. Adopting this interpretation, fuzzy set theory has become a basic mathematical model to represent *imprecision* and *vagueness*, but of course many other different interpretations are also possible. A typical example which explains how fuzzy sets can be used in order to deal with imprecision is about the *height* of a person (we will turn back to this example in the last section of this chapter). Indeed, when we are asked whether an individual x belongs to the set of *tall* persons, the classical truth values 1 (true) and 0 (false) might be insufficient. On the other hand, values in the real unit interval $[0, 1]$ provide a wider spectrum with which one can evaluate to what extent an individual can be considered as tall. In this prospective the *fuzzy* set of tall persons becomes a function $\mu_{tall} : X \to [0, 1]$ from the set X of individuals to $[0, 1]$, assigning to each individual x its degree $\mu_{tall}(x)$ of being tall. We refer the reader e.g. to [20,32,39] for monographs on the topic.

In the literature several attempts to extend belief functions on fuzzy events can be found. The first extensions of Dempster-Shafer theory to the general framework of fuzzy set theory was proposed by Zadeh in the context of information granularity and possibility theory [59] in the form of an expected conditional necessity, and by Smets who proposed in [51] to extend a classical belief function Bel on 2^X to fuzzy subsets A of X as the lower expectation of the characteristic function of A with respect to the class of probability measures lower bounded by Bel. After Zadeh and Smets, several further generalizations were proposed depending on the way a measure of *inclusion among fuzzy sets* is used to define the belief functions of fuzzy events based on fuzzy evidence. Indeed, given a mass assignment m for the bodies of evidence $\{A_1, A_2, \ldots\}$, and a measure $I(A \subseteq B)$ of inclusion among fuzzy sets, the belief of a fuzzy set B can be defined in general by the value: $Bel(B) = \sum_i I(A_i \subseteq B) \cdot m(A_i)$. We refer the reader to [37,55,56] for exhaustive surveys, and to [2] for another approach through fuzzy subsethood.

Belief functions on distributive lattices were studied in [33] and [61] where the authors define, starting from a given mass assignment $m : L \to [0, 1]$, the belief degree $Bel(a)$ of an element a of a distributive lattice L to be $Bel(a) = \sum_{x \in I_L, x \leq a} m(x)$, where I_L denotes the set of all join-irreducible elements of L. Notice that, although the framework of distributive lattices is much more general than the framework we are going to discuss in this chapter (we invite the reader to consult Section 3), the inclusion operator used in [33,61] is crisp and hence it does not take into account a graded notion of inclusion.

Different definitions were also introduced by Dubois and Prade [19] and by Denœux [15,16] to deal with belief functions ranging over intervals or fuzzy numbers.

Of course, moving from classical sets to fuzzy sets conveys non trivial complications in the description of the algebraic model aimed at representing the available evidence. In particular, although we start with a finite set X, while the Boolean algebra 2^X is finite and hence atomic, the class $[0,1]^X$ of the fuzzy subsets $f : X \to [0,1]$ of X contains uncountably many elements and hence it is not always trivial to define a mass m over them. Moreover there are several different ways to generalize Boolean algebras to algebras of fuzzy sets. Usually the generalization of belief functions to this frame, is done in the algebraic context of the so called *De Morgan triples* (or *Zadeh-algebras*) over classes of fuzzy sets, and where intersection, union, and complementation, are replaced in $[0,1]^X$ by the pointwise extensions of the operations in $[0,1]$ of min, max, and standard negation $\neg : x \mapsto 1 - x$ respectively (see for instance [60, §2.1]).

In [25,41,42], to further generalize belief functions on fuzzy sets, the authors frame their investigation in the algebraic setting of MV-algebras [7,45] (in fact in every MV-algebra a Zadeh-algebra is obtainable as a reduct) and, since a belief function can be equivalently represented by a probability measure $P_m : 2^{2^X} \to [0,1]$ such that $P(\{\emptyset\}) = 0$, they replace the usual probability measures by the notion of *state* on MV-algebras [44]. Indeed MV-algebras are the algebraic structures for fuzzy sets enabling the most natural treatment of many-valued probability theory. The reason is in the formula for a probability of a fuzzy event proposed already by Zadeh [58]. His definition — the expected value of the membership function of the fuzzy set w.r.t. a probability measure on its domain — later turned out to be a consequence of the axiomatic treatment of MV-probability. We will provide a more detailed introduction about these topics in Subsection 3.1.

In this chapter we will survey recent developments on belief functions on MV-algebras of fuzzy sets, mainly following the lines of the already above cited papers [25,28,41,42]. The paper is organized as follows. In Section 2 we recall how belief functions are defined on Boolean algebras and in particular we will present a first definition based on mass assignments, and a second (equivalent) one based on probability measures. Then, in Section 3 we introduce the main algebraic structures we will need along the paper, namely MV-algebras. In the same section we also introduce states on MV-algebra (in Subsection 3.1) and we recall some basic results we are going to use later. Section 4 contains two main approaches to define belief functions on MV-algebras: we will deal with a first approach in which belief functions on fuzzy sets are built up over crisp, Boolean focal elements (Subsection 4.1), and a second, more general approach, in which belief functions on fuzzy sets are defined in a way to allow for focal elements to be fuzzy as well (Subsection 4.2). Belief functions on MV-algebras are not necessarily *normalized* measures, in the sense of the belief of the empty set being zero. We will discuss the normalization problem in Section 5 and the case of a *soft-normalization* of mass assignments, and hence of belief functions, in Subsection 6. Then in Section 7 we present a generalization of Dempster rule of combination, we discuss some particular cases and we provide an example aimed at clarifying the use of such a construction in the general frame of fuzzy sets. We will end this chapter with Section 8 where we present some concluding remarks and we also suggest alternative readings about the topic.

2 Belief Functions on Boolean Algebras

Consider a finite set X whose elements can be regarded as mutually exclusive (and exhaustive) propositions of interest, and whose powerset 2^X represents all combinations of such propositions. The set X is usually called the *frame of discernment*, and every element $x \in X$ represents the lowest level of discernible information we can deal with.

A map $m : 2^X \to [0, 1]$ is said to be a *basic belief assignment*, or a *mass assignment* whenever

$$m(\emptyset) = 0 \text{ and } \sum_{A \in 2^X} m(A) = 1.$$

Given such a mass assignment m on 2^X, for every $A \in 2^X$, the *belief of A* is defined as

$$Bel_m(A) = \sum_{B \subseteq A} m(B). \tag{1}$$

Every mass assignment m on 2^X is in fact a probability distribution on 2^X that naturally induces a probability measure P_m on 2^{2^X}. Consequently, the belief function Bel_m corresponding to m can be equivalently described as follows: for every $A \in 2^X$,

$$Bel_m(A) = P_m(\{B \in 2^X : B \subseteq A\}). \tag{2}$$

Therefore, identifying the set $\{B \in 2^X : B \subseteq A\}$ with its characteristic function on 2^{2^X} defined by

$$\beta_A : B \in 2^X \mapsto \begin{cases} 1 & \text{if } B \subseteq A, \\ 0 & \text{otherwise,} \end{cases} \tag{3}$$

it is easy to see that, for every $A \in 2^X$, and for every mass assignment $m : 2^X \to [0, 1]$, we have

$$Bel_m(A) = P_m(\beta_A). \tag{4}$$

This easy characterization will be important when we discuss the extensions of belief functions on MV-algebras. The following is a trivial observation about the map β_A that can be useful to understand our generalization: for every $A \in 2^X$, β_A can be regarded as a map evaluating the (Boolean) inclusion of B into A, for every subset B of X.

A subset A of X such that $m(A) > 0$ is said to be a *focal element*. Every belief function is characterized by the value that m takes over its focal elements, and therefore, the focal elements of a belief function Bel_m contain the pieces of evidence that characterize Bel_m itself. For every set X and for every mass assignment m, call $\mathfrak{F}_m$ the set of focal elements of 2^X with respect to m. It is well known that several subclasses of belief functions can be characterized just by the structure of their focal elements. In particular, when $\mathfrak{F}_m \subseteq \{\{x\} : x \in X\}$, it is clear that Bel_m is indeed a probability measure. Moreover, if the focal elements are nested subsets of X, i.e. $\mathfrak{F}_m$ is a chain with respect to the inclusion relation between sets, then Bel_m is a *necessity measure* [19,50]; this means e.g. that in such a case, it holds that $Bel_m(A_1 \cap A_2) = \min\{Bel_m(A_1), Bel_m(A_2)\}$ for eveary $A_1, A_2 \in 2^X$.

The whole class of belief functions on Boolean algebras is characterized by the property of non-decreasing differences of all possible orders. This property can be formulated for any function $v : L \to \mathbb{R}$ defined on a distributive lattice L [33,61]. We say that v is *totally monotone* if, for every $n \geq 2$ and every $a_1, \ldots, a_n \in L$, we have

$$v\left(\bigvee_{i=1}^{n} a_i\right) \geq \sum_{\emptyset \neq I \subseteq \{1,\ldots,n\}} (-1)^{|I|+1} v\left(\bigwedge_{i \in I} a_i\right).$$

Shafer [50] has shown that the following assertions are equivalent for a function $v : 2^X \to [0,1]$ such that $v(\emptyset) = 0$ and $v(X) = 1$:

- v is a belief function,
- v is a totally monotone function on the distributive lattice 2^X.

As we will see in the following sections, this property does not characterize, in general, belief functions on fuzzy sets.

3 MV-Algebras: An Algebraic Frame for Many-Valued Events

In the same way Boolean algebras are the algebraic structures related to classical logic, MV-algebras are the algebras naturally associated to infinitely-valued Łukasiewicz logic.

The language of Łukasiewicz logic Ł (cf. [7,35]), consists of a countable set of propositional variables $\{p_1, p_2, \ldots\}$, the binary connective $\to$, and the truth constant $\bot$. Formulas are defined by the usual inductive clauses. The following formulas provide an axiomatization for Ł:

(**Ł**1) $\varphi \to (\psi \to \varphi)$
(**Ł**2) $(\varphi \to \psi) \to ((\psi \to \chi) \to (\varphi \to \chi))$
(**Ł**3) $((\varphi \to \bot) \to (\psi \to \bot)) \to (\psi \to \varphi)$
(**Ł**4) $((\varphi \to \psi) \to \psi) \to ((\psi \to \varphi) \to \varphi)$

The rule of inference of Ł is *modus ponens*: from φ and $\varphi \to \psi$, deduce ψ.

Further connectives in Ł are definable as follows: $\neg\varphi = \varphi \to \bot$; $\varphi \oplus \psi = \neg\varphi \to \psi$; $\varphi \odot \psi = \neg(\varphi \to \neg\psi)$; $\varphi \vee \psi = (\varphi \to \psi) \to \psi$; $\varphi \wedge \psi = \neg(\neg\varphi \vee \neg\psi)$; $\top = \neg\bot$.

Łukasiewicz logic is an algebraizable logic in the sense of Blok and Pigozzi [3], and its equivalent algebraic semantics is constituted by the class of MV-algebras [6,7]. In algebraic terms, an MV-algebra is a structure $\mathbf{A} = (A, \oplus, \neg, 0)$ of type $(2,1,0)$ satisfying the following equations, for every $a, b, c \in A$:

(MV1) $a \oplus (b \oplus c) = (a \oplus b) \oplus c$,
(MV2) $a \oplus b = b \oplus a$,
(MV3) $a \oplus 0 = a$,
(MV4) $\neg\neg a = a$,
(MV5) $a \oplus \neg 0 = \neg 0$,
(MV6) $\neg(\neg a \oplus b) \oplus b = \neg(\neg b \oplus a) \oplus a$.

Further (definable) operations can be defined from $\oplus, \neg$ and 0 in a similar way as for the logical connectives above. In particular: $a \to b = \neg a \oplus b$, $a \odot b = \neg(\neg a \oplus \neg b)$; $a \vee b = \neg(\neg a \oplus b) \oplus b$; $a \wedge b = \neg(\neg a \vee \neg b)$; $1 = \neg 0$.

For every MV-algebra $\mathbf{A} = (A, \oplus, \neg, 0, 1)$, the structure $\mathbf{L}(\mathbf{A}) = (A, \wedge, \vee, 0, 1)$, where $\wedge$ and $\vee$ are defined as above, is a bounded distributive lattice and moreover the order relation $\leq$ defined by the stipulation: for all $a, b \in A$

$$x \leq y \text{ iff } x \to y = 1,$$

coincides with the lattice order of $\mathbf{L}(\mathbf{A})$. An MV-algebra whose order $\leq$ is linear is called an *MV-chain.* The class of MV-algebras forms a variety that we denote by $\mathbb{MV}$.

Let $\mathbf{A}$ be an MV-algebra. Then a non empty subset $\mathfrak{f}$ of A is said to be a *filter* of $\mathbf{A}$ iff: (i) $1 \in \mathfrak{f}$, (ii) if $a, b \in \mathfrak{f}$, then $a \odot b \in \mathfrak{f}$, and (iii) if $a \in \mathfrak{f}$ and $b \geq a$, then $b \in \mathfrak{f}$. A filter $\mathfrak{f}$ of an MV-algebra $\mathbf{A}$ is said to be *proper*, if $\mathfrak{f} \neq A$. A filter $\mathfrak{m}$ is said to be a *maximal filter* (or an *ultrafilter*) whenever for any proper filter $\mathfrak{f}$ such that $\mathfrak{f} \supseteq \mathfrak{m}$, either $\mathfrak{f} = A$, or $\mathfrak{f} = \mathfrak{m}$. The set of all ultrafilters of an MV-algebra $\mathbf{A}$ will be henceforth denoted by $\mathfrak{M}(\mathbf{A})$, or, when there is no danger of confusion, simply by $\mathfrak{M}$. For every MV-algebra $\mathbf{A}$, the set $\mathfrak{M}(\mathbf{A})$ is non-empty and it can be endowed with a compact Hausdorff topology, the so-called *spectral topology*: for an arbitrary filter $\mathfrak{f}$ of A, any set of the form $O_{\mathfrak{f}} = \{\mathfrak{m} \in \mathfrak{M}(\mathbf{A}) : \mathfrak{m} \not\supseteq \mathfrak{f}\}$ is open in this topology.

Observe that an intersection of a family of filters is a filter. The intersection of the family of all maximal filters of an MV-algebra $\mathbf{A}$ is called the *radical* of $\mathbf{A}$ and it is usually written $Rad(\mathbf{A})$. An MV-algebra $\mathbf{A}$ is *semisimple* whenever $Rad(\mathbf{A}) = \{1\}$. It is well-known (see [7] for instance) that the congruences lattice and the filters lattice of any MV-algebra $\mathbf{A}$ are mutually isomorphic, via the isomorphism which associates to every congruence[1] θ the filter $\mathfrak{f}_\theta = \{a \in A \mid (a, 1) \in \theta\}$.

Example 3.1. The following are four relevant examples of MV-algebras:

(1) Every Boolean algebra is an MV-algebra, and moreover for every MV-algebra $\mathbf{A}$, the set $B(\mathbf{A}) = \{a \in A : a \oplus a = a\}$ of its idempotent elements is the domain of the largest Boolean subalgebra of $\mathbf{A}$. The algebra having $B(\mathbf{A})$ as universe is usually called the *Boolean skeleton* of $\mathbf{A}$.

[1] A congruence θ in a MV-algebra $\mathbf{A}$ is an equivalence relation on A respecting the operations, i.e. if $(x, y) \in \theta$ then $(\neg x, \neg y) \in \theta$, and if $(x, y), (x', y') \in \theta$ then $(x \oplus x', y \oplus y') \in \theta$.

(2) Define on the real unit interval $[0,1]$ the operations $\oplus$ and $\neg$ as follows: for all $a, b \in [0,1]$,

$$a \oplus b = \min\{1, a+b\}, \text{ and } \neg a = 1 - a.$$

Then the structure $[0,1]_{MV} = ([0,1], \oplus, \neg, 0)$ is an MV-algebra. The MV-algebra $[0,1]_{MV}$ is generic for the variety of MV-algebras (i.e. it generates the whole variety $\mathbb{MV}$) and it is usually called the *standard* MV-algebra. In equivalent terms, Łukasiewicz logic is complete with respect to the semantics defined by the standard MV-algebra.

(3) Fix $k \in \mathbb{N}$, and let $F(k)$ be the set of all the McNaughton functions (cf. [7]) from the hypercube $[0,1]^k$ into $[0,1]$. In other words, $F(k)$ is the set of all the functions $f : [0,1]^k \to [0,1]$ which are continuous, piecewise linear and such that each linear piece has integer coefficients only. The following pointwise operations defined on $F(k)$,

$$(f \oplus g)(x) = \min\{1, f(x) + g(x)\}, \text{ and } (\neg f)(x) = 1 - f(x),$$

make the structure $\mathbf{F}(k) = (F(k), \oplus, \neg, \overline{0})$ into an MV-algebra, where $\overline{0}$ clearly denotes the function constantly equal to 0. Actually, $\mathbf{F}(k)$ is the free MV-algebra over k generators [7].

(4) Let X be a non-empty set, and let $\mathbf{A} = [0,1]^X$ the set of all functions from X into $[0,1]$, endowed with operations defined by the pointwise application of those in $[0,1]_{MV}$. The structure $[0,1]^X$ is clearly an MV-algebra, which we will henceforth call *MV-algebra of fuzzy sets* in order to point out that every fuzzy subset of X is indeed included into A. Every MV-subalgebra of $[0,1]^X$ is called an *MV-clan* or simply a *clan* (cf. [4,46]). Notice that, for a finite non-empty set X, the Boolean skeleton of the MV-algebra of fuzzy sets $[0,1]^X$ coincides with the power set 2^X of X.

Notation 1. As already recalled in the introduction, MV-algebras, and MV-clans in particular, constitute the algebraic framework on top of which we will define belief functions. Indeed, elements of an MV-algebra $\mathbf{A}$ will be always intended to be the *fuzzy sets* we will work with. Therefore, although in the previous sections we used the notation $f, g, \ldots$ to indicate fuzzy sets, we will henceforth denote them by $a, b, c, \ldots$ without danger of confusion. At the same time, the notation $f, g, \ldots$ will be reserved to indicate functions in general. Moreover, in order to distinguish fuzzy sets from crisp sets, the latter will be indicated using capitals letters. So, for example, for any MV-algebra $\mathbf{A}$ we will denote the elements of $B(\mathbf{A})$ by $C, D, \ldots$, while for generic elements of $\mathbf{A}$ we will use the notation $a, b, c, \ldots$.

It is worth noticing that in $[0,1]_{MV}$, the interpretation of the lattice operations of $\wedge$ and $\vee$ is, respectively, in terms of the min and max operators. Therefore, we will henceforth use both the notations $\wedge$ and min, and $\vee$ and max, without danger of confusion.

Roughly speaking the class of MV-algebras can be divided into semisimple and non-semisimple MV-algebras. This is, in particular, a key point of distinction between MV and Boolean algebras. Remember, in fact, that every Boolean algebra is semisimple, and that all MV-algebras of fuzzy sets $\mathbf{A} = [0,1]^X$ are semisimple as well.

A semisimple MV-algebra $\mathbf{A} \subseteq [0,1]^X$ is said to be *separating* provided that for each $x_1 \neq x_2 \in X$, there is a $a \in A$ such that $a(x_1) \neq a(x_2)$. Hence, the MV-algebras of fuzzy sets $\mathbf{A} = [0,1]^X$ are both separating and semisimple. The following theorem provides a representation of semisimple MV-algebras by algebras of continuous functions.

Theorem 3.2 (Chang [6], Belluce [1]). *Up to isomorphism, every semisimple MV-algebra* $\mathbf{A}$ *is a separating algebra of continuous* $[0,1]$*-valued functions over the compact Hausdorff space* $\mathfrak{M}(\mathbf{A})$ *of ultrafilters of* $\mathbf{A}$.

The following result, which we state for the particular case of MV-algebras of fuzzy sets, holds in a much more general setting. We invite the interested reader to consult [7] for a more exhaustive treatment.

Theorem 3.3. *For every MV-algebra* $\mathbf{A} = [0,1]^X$*, there exists a one-to-one correspondence between the points of* X *and the class* $Hom(\mathbf{A}, [0,1]_{MV})$ *of homomorphisms of* $\mathbf{A}$ *into the standard MV-algebra* $[0,1]_{MV}$.

Thanks to the above Theorem 3.3 we will henceforth identify points in X with homomorphisms of $\mathbf{A}$ in the standard MV-algebra $[0,1]_{MV}$ without loss of generality. Moreover, the following holds:

Corollary 3.4. *Let* $\{\tau_1, \ldots, \tau_s\}$ *be a finite subset of an MV-algebra* $\mathbf{A} = [0,1]^X$. *Then*

$$\{\langle h(\tau_1), \ldots, h(\tau_s)\rangle \in [0,1]^s : h \in Hom(\mathbf{A}, [0,1]_{MV})\} =$$
$$\{\langle \tau_1(x), \ldots, \tau_s(x)\rangle : x \in X\}.$$

In this paper we will mainly concentrate on MV-algebras which are MV-clans $[0,1]^X$ defined over a *finite* set X. Such MV-algebras can be identified with finite direct products of copies of $[0,1]_{MV}$.

3.1 States on MV-Algebras

States on MV-algebras have been introduced by Mundici in [44] as averaging processes for the infinitely-valued Łukasiewicz calculus.

Definition 3.5. Let $\mathbf{A}$ be an MV-algebra. A *state* on $\mathbf{A}$ is a map $s : A \to [0,1]$ such that:

(s1) $s(1) = 1$,
(s2) For all $a, b \in A$ such that $a \odot b = 0$, $s(a \oplus b) = s(a) + s(b)$.

A state s on an MV-algebra $\mathbf{A}$ is said to be *faithful* if $s(x) = 0$ implies $x = 0$.

For a given MV-algebra $\mathbf{A}$, the class of all states on $\mathbf{A}$ is denoted by $\mathcal{S}(\mathbf{A})$. States play the same role on MV-algebra as probability measures do on Boolean algebras: indeed, the two properties (s1) and (s2) characterize each state on $\mathbf{A}$ as a $[0,1]$-valued map that is normalized (s1) and additive (s2) with respect to the MV-algebraic operations. Moreover, it is easy to see that, for every MV-algebra $\mathbf{A}$ and for every $s \in \mathcal{S}(\mathbf{A})$, the restriction of s to the Boolean skeleton $B(\mathbf{A})$ of $\mathbf{A}$ is a finitely additive probability measure. The following theorem, independently proved in [40] and [47], shows an integral representation of states by Borel probability measures defined on the σ-algebra $\mathfrak{B}(X)$ of Borel subsets of X, where X is any compact Hausdorff topological space.

Theorem 3.6. *Let $\mathbf{A} \subseteq [0,1]^X$ be a separating clan of continuous functions over a compact Hausdorff space X. Then there is a one-to-one correspondence between the class $\mathcal{S}(\mathbf{A})$ of states on $\mathbf{A}$, and the regular Borel probability measures on $\mathfrak{B}(X)$. In particular, for every state s on $\mathbf{A}$, there exists a unique regular Borel probability measure μ on $\mathfrak{B}(X)$ such that for every $a \in A$,*

$$s(a) = \int_X a \, \mathrm{d}\mu. \tag{5}$$

In the next example we consider a particular case of states on MV-algebras of fuzzy sets, focusing on the integral representation presented above.

Example 3.7. Let X be a finite non empty set. Let $\mathbf{A} = [0,1]^{(2^X)}$ be the MV-algebra of fuzzy sets consisting of all functions from 2^X to $[0,1]$ (i.e. $\mathbf{A}$ is the MV-algebra of all fuzzy subsets of the powerset 2^X of X). We will henceforth deal with those states on $[0,1]^{(2^X)}$ satisfying $s(\chi_{\{\emptyset\}}) = 0$ (where $\chi_{\{\emptyset\}}$ denotes the characteristic function of $\emptyset \in 2^X$). The above Theorem 3.6 ensures that for each such state s, there exists a unique finitely additive probability measure $\mu : 2^{(2^X)} \to [0,1]$ such that, for every $a \in [0,1]^{(2^X)}$,

$$s(a) = \sum_{A \subseteq X} a(A) \cdot \mu(\{A\}),$$

and $\mu(\{\emptyset\}) = 0$.

Obviously the class $\mathcal{S}(\mathbf{A})$ is non-empty since $Hom(\mathbf{A}, [0,1]_{MV}) \subseteq \mathcal{S}(\mathbf{A})$. Moreover $\mathcal{S}(\mathbf{A})$ is a convex subset of the compact Hausdorff space $[0,1]^A$ whose set of extremal points coincides with $Hom(\mathbf{A}, [0,1]_{MV})$. For every subset X of a topological vector space, let us write $\overline{\mathrm{co}}(X)$ to denote the closure of the convex hull of the set X [21]. Then, Krein-Mil'man Theorem [31] gives the following result.

Theorem 3.8. *For every MV-algebra $\mathbf{A}$, $\mathcal{S}(\mathbf{A}) = \overline{\mathrm{co}}(Hom(\mathbf{A}, [0,1]_{MV}))$.*

The following example is obtained by applying the above Theorem 3.8 to the particular case of MV-algebras of fuzzy sets, and it will be needed in the remaining of part of this chapter.

Example 3.9. As in Example 3.7, let X be finite and let $\mathcal{S}_0$ be the subset of $\mathcal{S}([0,1]^{(2^X)})$ of those states $s : [0,1]^{(2^X)} \to [0,1]$ further satisfying $s(\chi_{\{\emptyset\}}) = 0$.

The set $\mathcal{S}_0$ is a convex subset of the $(2^{|X|}-1)$-dimensional Euclidean space. Since the correspondence between $\mathcal{S}_0$ and the set of all probabilities μ on $2^{(2^X)}$ satisfying $\mu(\{\emptyset\})=0$ is a one-to-one affine mapping, the set $\mathcal{S}_0$ is a $(2^{|X|}-2)$-simplex as well. Regarding the extreme points of $\mathcal{S}_0$, we can observe that they are in one-to-one correspondence with the non-empty subsets of X, and hence every state s_A, with $A \in 2^X \setminus \{\emptyset\}$, such that $s_A(a)=a(A)$ for each $a \in [0,1]^{(2^X)}$ is an extreme point of $\mathcal{S}_0$.

4 Belief Functions on MV-Algebras of Fuzzy Sets

In this section we are going to discuss two main MV-algebraic generalizations of belief functions. Our approach is to generalize the definition (4), where Bel_m is derived from a probability measure P_m on 2^{2^X}. Therefore, we need to generalize both the inclusion map β_A and the probability measure P_m. In the following subsections we investigate two directions in which these notions can be generalized.

4.1 The Case of Crisp Focal Elements

Let X be a finite nonempty set, and let, for each element a in the MV-algebra $[0,1]^X$, the map $\hat{\rho}_a : 2^X \to [0,1]$ be defined by the following stipulation: for all $B \in 2^X$,

$$\hat{\rho}_a(B) = \begin{cases} \min_{x\in B} a(x) & \text{if } B \neq \emptyset, \\ 1 & \text{if } B = \emptyset. \end{cases} \tag{6}$$

Remark 4.1. $\hat{\rho}_a$ generalizes the map β_A we discussed in Section 2 in the following sense: whenever $A \in B(\mathbf{A}) = 2^X$, then $\hat{\rho}_A = \beta_A$. Indeed, for every $A \in B(\mathbf{A})$, $\hat{\rho}_A(B) = 1$ if $B \subseteq A$, and $\hat{\rho}_A(B) = 0$, otherwise.

Definition 4.2 (Crisp Focal Elements). Let X be a finite nonempty set. Then a map $\hat{\mathbf{b}} : [0,1]^X \to [0,1]$ is a *crisp-focal element belief function*, if there exists a state $s : [0,1]^{(2^X)} \to [0,1]$ such that, for all $a \in [0,1]^X$

$$\hat{\mathbf{b}}(a) = s(\hat{\rho}_a).$$

The state s defining $\hat{\mathbf{b}}$ will be henceforth called the *state assignment* of $\hat{\mathbf{b}}$.

The integral representation theorem for states (Theorem 3.6) can be generalized to crisp-focal belief functions. This requires the introduction of the Choquet integral (cf. [14]). Let f be any function from a finite nonempty set X to $[0,1]$, and let σ be a set function $\sigma : 2^X \to [0,1]$ such that $\sigma(\emptyset)=0$. Then the *Choquet integral* of f with respect to σ is defined as

$$\oint f \,\mathrm{d}\sigma = \int_0^1 \sigma(f^{-1}([t,1]))\,\mathrm{d}t.$$

Since X is finite, the integral $\oint f \, \mathrm{d}\sigma$ exists and takes the form of a finite sum. In fact, without loss of generality let $X = \{x_1, \ldots, x_n\}$, where the numbers $y_i = f(x_i)$ satisfy $y_1 \geq \cdots \geq y_n$. Put $y_{n+1} = 0$ and for each $i = 1, \ldots, n$, $S_i = \{x_1, \ldots, x_i\}$, then

$$\oint f \, \mathrm{d}\sigma = \sum_{i=1}^{n} (y_i - y_{i+1})\sigma(S_i).$$

Proposition 4.3. *For every crisp-focal belief function* $\hat{\mathbf{b}} : [0,1]^X \to [0,1]$ *there exists a unique belief function* $Bel : 2^X \to [0,1]$ *such that, for each* $a \in [0,1]^X$,

$$\hat{\mathbf{b}}(a) = \oint a \, \mathrm{d}(Bel).$$

Proof. Let s be the state assignment on $[0,1]^{(2^X)}$ which defines $\hat{\mathbf{b}}$. According to Example 3.7 there is a unique finitely additive probability measure μ on $2^{(2^X)}$ such that, for each $f \in [0,1]^{(2^X)}$, one has $s(f) = \sum_{A \subseteq X} f(A) \cdot \mu(\{A\})$ and $\mu(\{\emptyset\}) = 0$. Therefore, the crisp-focal belief function $\hat{\mathbf{b}}$ is expressed as follows: for every $a \in [0,1]^X$,

$$\hat{\mathbf{b}}(a) = s(\hat{\rho}_a) = \sum_{A \subseteq X} \hat{\rho}_a(A) \cdot \mu(\{A\}). \tag{7}$$

Recalling the definition (3) of the map β_A, we have $\hat{\rho}_a(A) = \min\{a(x) : x \in A\} = \oint a \, \mathrm{d}\beta_A$, for every $a \in [0,1]^X$ and for every $A \subseteq X$. Equation (7) together with the linearity of Choquet integral with respect to the integrating set function β_A yields

$$\hat{\mathbf{b}}(a) = \sum_{A \in 2^X \setminus \{\emptyset\}} \mu(\{A\}) \cdot \oint a \, \mathrm{d}\beta_A = \oint a \, \mathrm{d}\left(\sum_{A \in 2^X \setminus \{\emptyset\}} \mu(\{A\}) \cdot \beta_A \right).$$

The claim then follows noticing that the function $Bel : 2^X \to [0,1]$ such that for each $B \subseteq X$,

$$Bel(B) = \sum_{A \in 2^X \setminus \{\emptyset\}} \mu(\{A\}) \cdot \beta_A(B) = \mu(\{A \subseteq X \mid A \subseteq B\})$$

is a belief function on 2^X. □

With X being finite, despite the previous representation theorem for crisp-focal belief functions in terms of Choquet integral, Theorem 3.6 and Example 3.7 yields a unique probability measure $\mu : 2^{(2^X)} \to [0,1]$ such that for every $a \in [0,1]^X$

$$s(\hat{\rho}_a) = \sum_{C \in 2^X} \hat{\rho}_a(C) \cdot \mu(\{C\}). \tag{8}$$

Moreover, it is easy to see that, for every $C \subseteq 2^X$, $\mu(\{C\}) = s(\{C\})$, where $s(\{C\})$ is a succint expression for $s(\chi_{\{C\}})$. Since $\mu(\{\emptyset\}) = 0$, the probability measure μ induces a mass assignment m such that $m(C) = \mu(\{C\})$. This remark explains the name *crisp-focal* for the belief functions as in Definition 4.2. In fact, from (8), each crisp-focal belief function $\hat{\mathbf{b}}$ assigns, to each element $a \in [0,1]^X$, the value $\hat{\mathbf{b}}(a) = \sum_{C \subseteq X} \hat{\rho}_a(C) \cdot m(C)$. Therefore $\hat{\mathbf{b}}(a)$ is only determined by the *crisp* elements $C \subseteq X$ for which $m(C) > 0$, i.e. Boolean (crisp) focal elements.

In Dempster-Shafer theory, given a belief function $Bel : 2^X \to [0,1]$, the mass m that defines Bel can be recovered from Bel by the *Möbius transform* [50] of Bel:

$$m(A) = \sum_{B \subseteq A} (-1)^{|A \setminus B|} Bel(B).$$

In case of crisp-focal belief functions, the situation is analogous.

Proposition 4.4. *Let* $\hat{\mathbf{b}} : [0,1]^X \to [0,1]$ *be a crisp-focal belief function, defined as* $\hat{\mathbf{b}}(a) = s(\rho_a)$ *for some state* s *on* $[0,1]^{(2^X)}$ *such that* $s(\{\emptyset\}) = 0$ *and* $s(\{C\}) > 0$ *iff* $C(x) \in \{0,1\}$, *where* $C \neq \emptyset$. *Then*

$$s(\{A\}) = m(A) = \textstyle\sum_{B \subseteq A} (-1)^{|A \setminus B|} \hat{\mathbf{b}}(B)$$

for each $A \subseteq X$.

Proof. Definition (4.2) directly gives that $\hat{\rho}_A(C) \in \{0,1\}$ for each pair of crisp sets $A, C \subseteq X$ and thus

$$\hat{\mathbf{b}}(A) = \textstyle\sum_{C \in 2^X} \hat{\rho}_A(C) \cdot s(\{C\}) = \sum_{B \subseteq A} s(\{B\}) = \sum_{B \subseteq A} m(B).$$

This implies that the restriction of $\hat{\mathbf{b}}$ to 2^X is a classical belief function. See [42] for further details. □

As a corollary, observe that, in the hypothesis of the above proposition, the values $\hat{\mathbf{b}}(a)$ for non-crisp $a \in [0,1]^X$ are fully determined by the values of $\hat{\mathbf{b}}$ over the crisp sets of 2^X. Indeed, Proposition 4.3 proves that, for any $a \in [0,1]^X$, $\hat{\mathbf{b}}(a)$ is the Choquet integral of a with respect to the restriction of $\hat{\mathbf{b}}$ over 2^X. In this way we arrive at another characterization of crisp-focal belief functions.

Theorem 4.5. *A function* $\hat{\mathbf{b}} : [0,1]^X \to [0,1]$ *is a crisp-focal belief function iff its restriction onto* 2^X *is a totally monotone function, i.e., for every natural* n *and every* $A_1, \ldots, A_n \in 2^X$, *the following inequality holds:*

$$\hat{\mathbf{b}}\left(\bigvee_{i=1}^{n} A_i\right) \geq \sum_{\emptyset \neq I \subseteq \{1,\ldots,n\}} (-1)^{|I|+1} \cdot \hat{\mathbf{b}}\left(\bigwedge_{k \in I} A_k\right).$$

The geometrical structure of the set of all crisp-focal belief functions on $[0,1]^X$ is completely determined by the associated simplex of state assignments on $[0,1]^{(2^X)}$. For each $A \subseteq X$, a crisp focal belief function $\hat{\mathbf{b}}_A(a) = \min\{a(x) : x \in A\}$ for $a \in [0,1]^X$ corresponds to the state assignment s_A (see Example 3.9). Consequently, we obtain the following characterization of the set of all crisp focal belief functions.

Proposition 4.6. *The set of all crisp focal belief functions on* $[0,1]^X$ *is a* $(2^{|X|}-2)$*-simplex whose set of extreme points is* $\{\hat{\mathbf{b}}_A \mid A \in 2^X \setminus \{\emptyset\}\}$.

4.2 The Case of Fuzzy-Focal Elements

The notion of belief function on MV-algebras we are going to introduce in this section (cf. [25]) generalizes crisp-focal belief function by introducing, for every $a \in \mathbf{A} = [0,1]^X$, a more general inclusion map ρ_a associating with each *fuzzy set* $b \in A$ the degree of inclusion of the fuzzy set b into the fuzzy set a as follows:

$$\rho_a(b) = \min\{b(x) \Rightarrow a(x) : x \in X\}, \tag{9}$$

where $\Rightarrow$ is Lukasiewicz implication in the standard algebra $[0,1]_{MV}$ defined as $u \Rightarrow v = (\neg u) \oplus v = \min(1, 1-u+v)$, for all $u, v \in [0,1]$. The choice of $\Rightarrow$ in the above definition is clearly due to the MV-algebraic setting, but different choices could be made in other fuzzy logic settings.

The mapping ρ_a can be indeed regarded as a *generalized inclusion operator* between fuzzy sets (cf. [25] for further details) since the following intuitive properties are satisfied by such mappings:

- $\rho_a(b) = 1$ iff $b(x) \leq a(x)$, for all $x \in X$;
- $\rho_a(b) \geq \rho_a(b')$, whenever $b(x) \leq b'(x)$, for all $x \in X$;
- $\rho_a(b) = 0$ iff there is $x \in X$ such that $b(x) = 1$ and $a(x) = 0$.

Next proposition shows that the mapping ρ_a generalizes both the previously introduced mappings β_A and $\hat{\rho}_a$.

Proposition 4.7. (i) *For all* $a, a' \in \mathbf{A}$, $\rho_{a\wedge a'} = \min\{\rho_a, \rho_{a'}\}$, *and* $\rho_{a\vee a'} \geq \max\{\rho_a, \rho_{a'}\}$.

(ii) *For every* $a \in \mathbf{A}$, *the restriction of* ρ_a *to* $B(\mathbf{A})$ *coincides with the transformation* $\hat{\rho}_a$ *defined by* (*6*).

(iii) *For every* $A \in B(\mathbf{A})$, *the restriction of* ρ_A *to* $B(\mathbf{A})$ *coincides with the transformation* β_A *defined by* (*3*).

Proof. (i) In every MV-chain, and in particular in the standard chain $[0,1]_{\mathrm{MV}}$ the equation $\neg c \oplus (a \wedge b) = (\neg c \oplus a) \wedge (\neg c \oplus b)$ holds:, i.e. $(c \Rightarrow (a \wedge b)) = (c \Rightarrow a) \wedge (c \Rightarrow b)$. Therefore, for every $a, a', b \in \mathbf{A}$,

$$\begin{aligned}
\rho_{a\wedge a'}(b) &= \min\{b(x) \Rightarrow (a \wedge a')(x) : x \in X\} \\
&= \min\{b(x) \Rightarrow (a(x) \wedge a'(x)) : x \in X\} \\
&= \min\{(b(x) \Rightarrow a(x)) \wedge (b(x) \Rightarrow a'(x)) : x \in X\} \\
&= \min\{\rho_a(b), \rho_{a'}(b)\}.
\end{aligned}$$

An easy computation shows that $\rho_{a\vee a'} \geq \max\{\rho_a, \rho_{a'}\}$.

(ii) For every $B \in B(\mathbf{A})$, $\rho_a(B) = \min\{B(x) \Rightarrow a(x) : x \in X\}$. Whenever $x \notin B$, $B(x) = 0$, and hence $B(x) \Rightarrow a(x) = 1$ for all those $x \notin B$. On the other hand for all $x \in B$, $B(x) = 1$, and so $B(x) \Rightarrow a(x) = 1 \Rightarrow a(x) = a(x)$ for all $x \in B$. Consequently, $\rho_a(B) = \min\{a(x) : x \in B\}$.

(iii) It follows immediately from (ii) together with Remark 4.1. □

For every $A \in 2^X$ (i.e. whenever A is identified with a vector in $[0,1]^X$ having *integer* coordinates), the map $\rho_A : [0,1]^X \to [0,1]$ is a pointwise minimum of finitely many linear functions with integer coefficients, and hence ρ_A is a non-increasing McNaughton function [7].

Lemma 4.8. *The MV-algebra* $\mathbf{R_2}$ *generated by the set* $\varrho_2 = \{\rho_A : A \in 2^X\}$ *coincides with the free MV-algebra over* n *generators* $\mathbf{F}(n)$*, where* n *is the cardinality of* X*.*

Proof. By [8, Theorem 3.13], if a variety $\mathbb{V}$ of algebras is generated by an algebra $\mathbf{A}$, then the free algebra over a cardinal $n > 0$ is, up to isomorphisms, the subalgebra of $\mathbf{A}^{\mathbf{A}^X}$ generated by the projection functions $\theta_i : \mathbf{A}^X \to \mathbf{A}$. Therefore, in order to prove our claim it suffices to show that the projection functions $\theta_1, \ldots, \theta_n$ belong to ϱ_2.

For every $i = 1, \ldots, n$, let the vector $\bar{i} \in \{0,1\}^X$ be defined as

$$\bar{i}(j) = \begin{cases} 0, & \text{if } j = i \\ 1, & \text{otherwise.} \end{cases}$$

Then $\rho_{\bar{i}} = 1 - \theta_i$. In fact, for every $b \in [0,1]^X$, and for every $i, j \in X$ such that $j \neq i$, we have $b(j) \to \bar{i}(j) = 1$, and $b(i) \to \bar{i}(i) = 1 - b(i)$, so that $1 - \rho_{\bar{i}(b)} = \theta_i(b) = b(i)$. This actually shows that the MV-algebra $\mathbf{R_2^{\neg}}$ generated by the set $\neg\varrho_2 = \{1 - \rho_A : A \in 2^X\}$ is isomorphic to $\mathbf{F}(n)$. Clearly $\mathbf{R_2}$ and $\mathbf{R_2^{\neg}}$ are isomorphic through the map $g : a \in \mathbf{R_2} \mapsto 1 - a \in \mathbf{R_2^{\neg}}$. □

Therefore, if we consider the MV-algebra $\mathbf{R}_X$ generated by the set $\varrho = \{\rho_a : a \in [0,1]^X\}$ we obtain a semisimple MV-algebra that properly extends $\mathbf{F}(n)$, and whose elements are continuous functions from $[0,1]^X$ into $[0,1]$. This implies, in particular, that $\mathbf{R}_X$ is separating.

Definition 4.9 (Fuzzy-Focal Belief Function). Let X be a finite set and let $\mathbf{A} = [0,1]^X$. A map $\mathbf{b} : \mathbf{A} \to [0,1]$ will be called a *(fuzzy-focal) belief function* on the finite domain MV-clan $\mathbf{A}$ provided there exists a state $s : \mathbf{R}_X \to [0,1]$ such that for every $a \in A$,

$$\mathbf{b}(a) = s(\rho_a). \tag{10}$$

We will denote by $Bel(\mathbf{A})$ the class of all the (fuzzy-focal) belief functions over the finite domain MV-clan $\mathbf{A}$.

In analogy with the case of crisp-focal belief functions, the state s defining $\mathbf{b}$ will be henceforth called the *state assignment* of $\mathbf{b}$.

As in the previous section, we will identify the mass of a belief function $\mathbf{b}$ with the unique Borel regular probability measure μ over $\mathfrak{B}([0,1]^X)$ that represents the state s via Theorem 3.6.

Remark 4.10. Note that if the set $\{b \in [0,1]^X \mid \mu(\{b\}) > 0\}$ is countable then the above Definition 4.9 yields

$$\mathbf{b}(a) = \sum_{b \in A} \rho_a(b) \cdot \mu(\{b\}). \tag{11}$$

In this case, a *focal element* is any $b \in A$ such that $\mu(\{b\}) > 0$ and hence, in contrast with the case of crisp-focal belief functions, it is clear that focal elements of $\mathbf{b}$ can be proper fuzzy sets.

We showed in Theorem 4.5 that the property of total monotonicity characterizes crisp focal belief functions on MV-algebras. As for the case of fuzzy-focal belief functions, the problem of characterizing those belief functions in terms of (a variant of) total monotonicity is open, but the following implication holds.

Proposition 4.11. *For every finite-domain MV-clan* $\mathbf{A}$ *and for every* $\mathbf{b} \in Bel(M)$, $\mathbf{b}$ *is totally monotone on the lattice reduct of* $\mathbf{A}$.

Proof. Since for every $a \in \mathbf{A}$, ρ_a is monotone, and every state s is monotone, $\mathbf{b}$ is monotone as well. Moreover, for every n and for every $a_1, \ldots, a_n \in \mathbf{A}$, from (10) and Proposition 4.7 (i) we have the following chain of inequalities:

$$\begin{aligned}\mathbf{b}\left(\bigvee_{i=1}^n a_i\right) &= s(\rho_{a_1 \vee \ldots \vee a_n}) \\ &\geq s(\rho_{a_1} \vee \ldots \vee \rho_{a_n}) \\ &= \textstyle\sum_{\emptyset \neq I \subseteq \{1,\ldots,n\}} (-1)^{|I|+1} \cdot s\left(\bigwedge_{k \in I} \rho_{a_k}\right) \\ &= \textstyle\sum_{\emptyset \neq I \subseteq \{1,\ldots,n\}} (-1)^{|I|+1} \cdot s\left(\rho_{(\bigwedge_{k \in I} a_k)}\right) \\ &= \textstyle\sum_{\emptyset \neq I \subseteq \{1,\ldots,n\}} (-1)^{|I|+1} \cdot \mathbf{b}\left(\bigwedge_{k \in I} a_k\right).\end{aligned}$$

□

Since belief functions on $[0,1]^X$ are defined by states on $\mathbf{R}_X$ and different states s_1 and s_2 determine different belief functions $\mathbf{b}_1$ and $\mathbf{b}_2$, the set $Bel([0,1]^X)$ of belief functions on $[0,1]^X$ is in 1-1 correspondence with the set $\mathcal{S}(\mathbf{R}_X)$ of all states on $\mathbf{R}_X$. Moreover, this correspondence is an affine mapping. Hence $Bel([0,1]^X)$ is a compact convex subset of $[0,1]^{([0,1]^X)}$. Therefore Krein-Mil'man theorem shows that $Bel([0,1]^X)$ is in the closed convex hull of its extremal points. The following result characterizes the extremal points of $Bel([0,1]^X)$.

Proposition 4.12. *For every* $x \in [0,1]^X$, *the belief function* $\mathbf{b}_x$ *defined by*

$$\mathbf{b}_x(a) = s_x(\rho_a) = \rho_a(x), \quad a \in [0,1]^X, \tag{12}$$

is an extremal point of $Bel([0,1]^X)$.

Proof. A belief function $\mathbf{b} \in Bel([0,1]^X)$ is extremal iff its state assignment is extremal in $\mathcal{S}(\mathbf{R}_X)$. In fact s is not extremal iff there exist $s_1, s_2 \in \mathcal{S}(\mathbf{R}_X)$ and a real number $\lambda \in (0,1)$ such that $s = \lambda s_1 + (1-\lambda)s_2$. In particular, for every $a \in [0,1]^X$,

$$\mathbf{b}(a) = s(\rho_a) = \lambda s_1(\rho_a) + (1-\lambda)s_2(\rho_a) = \lambda \mathbf{b}_1(a) + (1-\lambda)\mathbf{b}_2(a),$$

whence $\mathbf{b}$ would not be extremal as well. □

As recalled above, $\mathbf{R}_X$ is separating. Therefore from Proposition 4.12 the extreme points of its state space are MV-homomorphisms s_x, for each $x \in [0,1]^X$. Hence the following holds due to (12).

Theorem 4.13. *Every belief function* $\mathbf{b}$ *is a pointwise limit of a convex combination of some functions* $\rho_{\cdot}(a^1),\ldots,\rho_{\cdot}(a^k)$*, where* $a^1,\ldots,a^k \in [0,1]^X$.

Remark 4.14. Consider the restriction $\mathbf{b}^-$ of a fuzzy-focal belief function $\mathbf{b}$ to the Boolean skeleton 2^X of its domain $[0,1]^X$. Then, although it has fuzzy-focal elements, the map $\mathbf{b}^-$ actually is a classical belief function since, by Proposition 4.11, $\mathbf{b}^-$ keeps being total monotone. Therefore, there exists a mass assignment on crisp subsets of X giving the same $\mathbf{b}^-$. In other words there exists a mass assignment $m^- : 2^X \to [0,1]$ such that, for every $A \in 2^X$ one has:

$$\mathbf{b}(A) = \mathbf{b}^-(A) = \sum_{B \subseteq A} m^-(B)$$

In the framework of finitely-valued fuzzy sets on the scale $S_k = \{0, 1/k, \ldots, (k-1)/k, 1\}$, an interesting question is how to compute the mass m^- from the mass μ giving $\mathbf{b}$, that is, what is the map $m^- : 2^X \to [0,1]$ fulfilling the constraints

$$\sum_{a \in (S_k)^X} \rho_A(a) \cdot \mu(a) = \sum_{B \subseteq A} m^-(B)$$

for each $A \in 2^X$. In fact, following [18,55], to find the solution the idea is to decompose the mass $\mu(a)$ of each fuzzy-focal element a into its level cuts a_{α_i}, with $\alpha_i \in S_k$, as follows:

$$m_a(a_{\alpha_i}) = \mu(a) \cdot (\alpha_i - \alpha_{i-1})$$

where $\alpha_i = i/k$, for $i = 0, 1, \ldots, k$. Finally, since it may be the case that two (or more) level sets of different fuzzy-focal elements coincide, we define for each $A \in 2^X$:

$$m^-(A) = \sum \{m_a(a_\alpha) \mid a \in (S_k)^X, \alpha \in S_k \text{ such that } a_\alpha = A\}.$$

5 On Normalized Belief Functions

The *height* of a fuzzy set $a \in [0,1]^X$ is defined in the literature as

$$h(a) = \max\{a(x) : x \in X\}. \tag{13}$$

The value $h(a)$ can be interpreted as the degree of normalization of a. As a matter of fact, a fuzzy set a is called *normalized* whenever $h(a) = 1$, otherwise it is called *non-normalized*. A non-normalized fuzzy set represents a partially inconsistent information.

The map $\rho_{\overline{0}}$ evaluating the degree of inclusion of any fuzzy set $b \in [0,1]^X$ in the empty fuzzy set $\overline{0}$ (constantly zero function) does not coincide, in general, with $\overline{0}$ itself. In fact, whenever b is a non-normalized fuzzy set (i.e. $h(b) < 1$), $\rho_{\overline{0}}(b) > 0$. Therefore, if s is a faithful state on $\mathbf{R}_X$, the fuzzy-focal belief function defined through s satisfies $\mathbf{b}(\overline{0}) > 0$.

Definition 5.1. A (fuzzy-focal) belief function $\mathbf{b} : [0,1]^X \to [0,1]$ is said to be *normalized* provided that

$$\mathbf{b}(\overline{0}) = s(\rho_{\overline{0}}) = 0. \tag{14}$$

In this section we will focus on normalized fuzzy-focal belief functions. Indeed it is worth noticing that crisp-focal belief functions are normalized, i.e. they always satisfy (14).

In classical Dempster-Shafer theory, the notion of focal element is crucial for classifying belief functions. Whenever $X = \{1, \ldots, n\}$ is a finite set, the Boolean algebra 2^X is finite, and hence the mass assignment $m : 2^X \to [0,1]$ obviously defines only finitely many focal elements. On the other hand, the MV-algebra $[0,1]^X$ has uncountably many elements, and hence we cannot find in general a mass assignment μ defined over $\mathfrak{B}([0,1]^X)$ and supported by an at most countable set only. This observation leads to the following definition.

Definition 5.2. Let $\mathcal{K}$ be the set of all compact subsets of an MV-algebra of fuzzy sets $[0,1]^X$. For every regular Borel probability measure μ defined on $\mathfrak{B}([0,1]^X)$, we call the set

$$\text{spt } \mu = \bigcap \{K | K \in \mathcal{K}, \mu(K) = 1\}$$

the *support of* μ.

By Theorem 3.6 we can regard spt μ as the support of the state s defined from μ via (5). In particular, the following holds:

$$\mathbf{b}(a) = \int_{[0,1]^X} \rho_a \, \mathrm{d}\mu = \int_{\text{spt } \mu} \rho_a \, \mathrm{d}\mu. \tag{15}$$

Therefore, for a belief function $\mathbf{b}$ on $[0,1]^X$ whose state assignment s is represented by a regular Borel probability measure μ, we will henceforth refer to spt μ as the set of focal elements of $\mathbf{b}$.

Proposition 5.3. *The set $\mathcal{S}_0$ of all states on $\mathbf{R}_X$ satisfying (14) is a nonempty compact convex subset of $[0,1]^{\mathbf{R}_X}$ considered with its product topology.*

Proof. $\mathcal{S}_0$ is nonempty: let s_1 be defined by

$$s_1(\rho) = \rho(\overline{1}),$$

for every $\rho \in \mathbf{R}_X$, where $\overline{1} : X \to [0,1]$ is the constant function of value 1. This gives in particular $s_1(\rho_{\overline{0}}) = \rho_{\overline{0}}(\overline{1}) = 0$ and thus $s_1 \in \mathcal{S}_0$. Let $s, s' \in \mathcal{S}_0$ and $\alpha \in (0,1)$. Then the function given by

$$\alpha s + (1-\alpha)s'$$

is a state on $\mathbf{R}_X$ which clearly satisfies (14). Hence $\mathcal{S}_0$ is a convex subset of the product space $[0,1]^{\mathbf{R}_X}$. Since the space $[0,1]^{\mathbf{R}_X}$ is compact, we only need to show that $\mathcal{S}_0$ is closed (in its subspace product topology). To this end, consider a convergent sequence $(s_m)_{m\in\mathbb{N}}$ in $\mathcal{S}_0$ whose limit is s. As the set of all states on $\mathbf{R}_X$ is closed, s is a state. That s satisfies (14) follows from the fact that $s(\rho_{\overline{0}}) = \lim_{m\to\infty} s_m(\rho_{\overline{0}}) = 0$. □

The family of states $\mathcal{S}_0$ can be characterized by employing the integral representation of states. Namely, we will show that a state assignment $s \in \mathcal{S}_0$ iff s is "supported" by normal fuzzy sets in $[0,1]^X$, i.e. fuzzy sets $a \in [0,1]^X$ such that $a(x) = 1$ for some $x \in X$. We will denote by $\mathcal{NF}(X)$ the set of normalized fuzzy sets from $[0,1]^X$, i.e.

$$\mathcal{NF}(X) = \{a \in [0,1]^X \mid a(x) = 1 \text{ for some } x \in X\}.$$

The following result characterizes normalized fuzzy-focal belief functions in terms of the support of their state assignment.

Theorem 5.4. *Let s be a state assignment on $\mathbf{R}_X$ and μ be the regular Borel probability measure associated with s. Then* spt $\mu \subseteq \mathcal{NF}(X)$ *if and only if $s \in \mathcal{S}_0$.*

Proof. Let μ be a probability measure on Borel subsets of $[0,1]^X$ such that spt $\mu \subseteq \mathcal{NF}(X)$. Put

$$s(a) = \int_{[0,1]^X} a \,\mathrm{d}\mu, \quad a \in \mathbf{R}_X. \tag{16}$$

Since $\rho_{\overline{0}}(a) = 0$ for each $a \in \text{spt } \mu$, it follows that

$$s(\rho_{\overline{0}}) = \int_{\text{spt } \mu} \rho_{\overline{0}} \,\mathrm{d}\mu = 0,$$

hence $s \in \mathcal{S}_0$. Conversely, assume that

$$s(\rho_{\overline{0}}) = \int_{[0,1]^X} \rho_{\overline{0}} \,\mathrm{d}\mu = 0,$$

which implies $\rho_{\overline{0}} = 0$ μ-almost everywhere over $[0,1]^X$. Since $\rho_{\overline{0}}(a) = 0$ iff $a \in [0,1]^X$ is such that $a(x) = 1$, for some $x \in X$, we obtain $\mu(\mathcal{NF}(X)) = 1$. □

In particular, every state assignment of a crisp-focal belief function belongs to $\mathcal{S}_0$.

6 Soft-Normalization for Fuzzy-Focal Belief Functions

Throughout the rest of the paper, we stipulate the following:

We always assume a mass μ such that its support spt μ *is countable.*

Consider a belief function $\mathbf{b}$ with a state assignment s supported by spt μ. Assume that there exists a focal element $a \in$ spt μ that is a non-normalized fuzzy set. Since spt μ is countable, we have neccesarily $\mu(\{a\}) > 0$,[2] and $\mathbf{b}$ is associating a positive degree of evidence to a (partially) inconsistent information, which is reflected on the value that $\mathbf{b}$ assigns to the empty fuzzy set $\overline{0}$. Indeed, in this case we have $\rho_{\overline{0}}(a) > 0$, and hence

$$\mathbf{b}(\overline{0}) = s(\rho_{\overline{0}}) = \sum_{b \in \text{spt } \mu} \rho_{\overline{0}}(b)\mu(\{b\}) \geq \rho_{\overline{0}}(a) \cdot \mu(\{a\}) > 0.$$

Notice that the more inconsistent the focal elements of $\mathbf{b}$ are, the greater is the value $\mathbf{b}(\overline{0})$. When events and focal elements are crisp sets (and hence the unique possible non-normalized focal element is $\overline{0}$), normalization consists in redistributing the mass that μ assigns to $\overline{0}$ to the other focal elements of μ (if any).

Dealing with fuzzy-focal elements makes it possible to introduce a notion of *soft normalization* for belief functions. In particular, this construction allows for a finer redistribution of the masses, which depends on two thresholds. Recall the notion of height $h(a)$ of a fuzzy set a introduced in (13). Then we introduce the following definition of α-normalization.

Definition 6.1. A mass assignment $\mu : [0,1]^X \to [0,1]$ is said to be *α-normalized* provided that $\inf\{h(a) : a \in \text{spt } \mu\} = \alpha$.

In other words, a mass is α-normalized provided that each focal element of μ has at least height α. In particular, for a belief function $\mathbf{b}$ we define the *degree of normalization* of $\mathbf{b}$ as the value

$$\inf\{h(a) : a \in \text{spt } (\mu)\},$$

where μ is the mass associated to $\mathbf{b}$.

Let now $\mu : [0,1]^X \to [0,1]$ be an α-normalized mass assignment, and assume that there exists a focal element b for m such that $h(b) = \beta > \alpha$.

The mass μ can be renormalized to the higher degree β by defining a new mass μ^β as follows: for every $a \in [0,1]^X$,

$$\mu^\beta(\{a\}) = \begin{cases} 0, & \text{if } h(a) < \beta \\ \frac{\mu(\{b\})}{1-K}, & \text{otherwise} \end{cases} \tag{17}$$

where $K = \sum_{h(l)<\beta} \mu(\{l\})$.

The idea of this *β-normalization*, similarly to the classical normalization, consists in fixing the value β as a new level of consistency for the mass we are considering. Since $\alpha < \beta \leq 1$, the class of focal elements of height lower then β is not empty. Then the process of β-normalization consists in redistributing all the mass which μ assigns to the fuzzy sets of height lower than β, i.e. $K = \sum_{h(l)<\beta} \mu(\{l\})$, to those focal elements of height greater of (or equal to) β.

Clearly, every mass μ can be renormalized up to a maximum value given by

[2] If spt μ is not countable, the condition $a \in$ spt μ does not guarantee $\mu(\{a\}) > 0$.

$$\beta_{max} = \sup\{h(a) : a \in \text{spt }(\mu)\}.$$

We will make use of β-normalization in the next section where we discuss a generalization of the Dempster rule of combination.

7 Generalized Dempster's Rule of Combination

The power of Dempster-Shafer theory is in the possibility of combining all the available evidences about an event. In order to describe this aggregation process, Dempster introduced in [13] the so called *Dempster rule of combination*, briefly recalled next. Given a frame of discernment X, consider two masses m_1 and m_2 on 2^X encoding the beliefs about evidences coming from two (possibly different) sources. Then the new mass assignment m on 2^X is obtained from m_1 and m_2 according to the Dempster rule is as follows: for every $Y \subseteq X$,

$$m(Y) = \sum_{A \cap B = Y} m_1(A) \cdot m_2(B) \tag{18}$$

This rule may result in a non-normalized mass assignment as soon as there exist focal elements A and B for m_1 and m_2 respectively such that $A \cap B = \emptyset$. The normalized version of the rule yields the mass assignment m' defined as $m'(\emptyset) = 0$ and for every $\emptyset \neq Y \subseteq X$,

$$m'(Y) = \frac{m(Y)}{1 - \left(\sum_{C \cap D = \emptyset} m_1(C) \cdot m_2(D) \right)}. \tag{19}$$

Let us illustrate this situation with the following well known example due to Smets [52] (see also [49]): Mr Jones has been murdered, and we know the murderer was in the set $X = \{Peter, Paul, Mary\}$. The only evidence we have is that Mrs Jones, who saw the killer leaving the scene of the murder, is 80% sure that the murderer is a man. Hence all the available evidence is expressed as $Prob(Man) = 0.8$.

As recalled, within Dempster-Shafer theory each piece of evidence is encoded by a mass assignment $m : 2^X \to [0, 1]$, assigning a value to each subset of X such that $\sum_{A \in 2^X} m(A) = 1$ and $m(\emptyset) = 0$ (i.e. there is no belief on the empty set).

Turning back to the case of Mr. Jones' murder, and since we know that $Prob(Man) = 0.8$, the set $\{Paul, Peter\}$ is a focal element of mass m_1, and in particular we assign $m_1(\{Paul, Peter\}) = 0.8$. We know nothing about the remaining probabilities, so we allocate the remaining mass 0.2 to the whole frame of discernment X (i.e. $m_1(X) = 0.2$). Therefore, the only focal elements in this example are $X = \{Peter, Paul, Mary\}$ and $\{Paul, Peter\}$.

Consider a possible second piece of evidence providing an alibi for Peter with confidence 0.6. This new information is hence encoded in the model with a new mass assignment m_2 such that $m_2(\{Paul, Mary\}) = 0.6$ and $m_2(X) = 0.4$.

The Dempster rule of combination given by (18) then provides a new mass assignment m resulting from the combination of m_1 and m_2:

$$\begin{aligned} m(\{\emptyset\}) &= 0; \\ m(\{Paul\}) &= 0.48; \\ m(\{Paul, Mary\}) &= 0.12; \\ m(\{Peter, Paul\}) &= 0.32; \\ m(\{X\}) &= 0.08. \end{aligned}$$

Notice that this mass is normalized and hence it coincides with the mass resulting from (19). From the combined mass assignment m we can compute the resulting belief function $\mathbf{b}_m : 2^X \to [0,1]$ as follows: for every $Y \subseteq X$

$$\mathbf{b}_m(Y) = \sum_{Z \subseteq Y} m(Z). \tag{20}$$

The previous formula yields for instance:

$$\begin{aligned} \mathbf{b}_m(\{Paul\}) &= 0.48; \\ \mathbf{b}_m(\{Paul, Mary\}) &= 0.6; \\ \mathbf{b}_m(\{Paul, Peter\}) &= 0.8; \\ \mathbf{b}_m(X) &= 1. \end{aligned}$$

In the example of Mr Jones' murder, the masses m_1 and m_2 were assigned to statements expressing precise properties regarding the individuals in the set of hypothesis X. On the other hand, we may argue that usually a description of the witness would be affected not only by uncertainty regarding the statements, but also by the imprecision of the statements. Therefore the classical framework would be insufficient to analyse the facts provided by the witness.

In [25] the authors present a generalization of the Dempster rule in order to combine the information carried by two belief functions $\mathbf{b}_1, \mathbf{b}_2 \in Bel([0,1]^X)$ into a single belief function $\mathbf{b}_{1,2} \in Bel([0,1]^X)$. In the rest of this section we will recall the basic steps of that construction and we will conclude with some remarks about the procedure. We start with an easy result about the definition of states in a product space needed in the construction of the generalized Dempster rule.

Proposition 7.1. *For every MV-algebra* $\mathbf{A}$, *and for every pair of states* $s_1, s_2 : \mathbf{A} \to [0,1]$, *there exists a state* $s^{1,2}$ *defined on the direct product* $\mathbf{A} \times \mathbf{A}$ *such that for every* $(b,c) \in \mathbf{A} \times \mathbf{A}$, $s^{1,2}(b,c) = s_1(b) \cdot s_2(c)$.

Let now $\mathbf{A} = [0,1]^X$, and let $\mathbf{R}_X$ be the MV-algebra defined in Section 4.2. Further, let s_1, s_2 be two state assignments on $\mathbf{R}_X$ such that $\mathbf{b}_1(a) = s_1(\rho_a)$ and $\mathbf{b}_2(a) = s_2(\rho_a)$ for all $a \in \mathbf{A}$. Assume that $\mu_1, \mu_2 : \mathfrak{B}(\mathbf{A}) \to [0,1]$ are the two regular probability measures of support spt μ_1 and spt μ_2, respectively, such that for $i = 1, 2$,

$$s_i(f) = \int_{\mathrm{spt}\ \mu_i} f \, \mathrm{d}\mu_i, \quad f \in \mathbf{R}_X.$$

Let

$$\mu_{1,2} : \mathfrak{B}(\mathbf{A} \times \mathbf{A}) \to [0,1]$$

be the product measure on Borel subsets generated by $\mathbf{A} \times \mathbf{A}$ and $s_{1,2}$ be the state on the MV-algebra of all measurable functions $\mathbf{A} \times \mathbf{A} \to [0,1]$ that is defined as an integral with respect to $\mu_{1,2}$.

Notice that $s_{1,2}$ actually coincides with the state $s^{1,2}$ on $\mathbf{R}_X \times \mathbf{R}_X$ defined as in Proposition 7.1 through the identification

$$s^{1,2}(g,h) = \int_{\mathfrak{B}(\mathbf{A})} g \, \mathrm{d}\mu_1 \cdot \int_{\mathfrak{B}(\mathbf{A})} h \, \mathrm{d}\mu_2 = \int_{\mathfrak{B}(\mathbf{A}\times\mathbf{A})} g \cdot h \, \mathrm{d}\mu_{1,2} = s_{1,2}(g \cdot h).$$

Hence, in particular, for any $g, h : \mathbf{A} \to [0,1]$ and f such that $f(x,y) = g(x) \cdot h(y)$, then Proposition 7.1 yields $s_{1,2}(f) = s_1(g) \cdot s_2(h)$.

Finally, for every $a \in \mathbf{A}$, consider the map $\rho_a^{\wedge} : \mathbf{A} \times \mathbf{A} \to [0,1]$ defined by

$$\rho_a^{\wedge}(b,c) = \rho_a(b \wedge c).$$

Then we are ready to define the following combination of belief functions.

Definition 7.2 (Generalized Dempster rule). Given $\mathbf{b}_1, \mathbf{b}_2 \in Bel(\mathbf{A})$ as above, define its min-*conjunctive combination* $\mathbf{b}_{1,2} : \mathbf{A} \to [0,1]$ as follows: for all $a \in \mathbf{A}$,

$$\mathbf{b}_{1,2}(a) = s_{1,2}(\rho_a^{\wedge}). \tag{21}$$

Regarding the support of the combined measure, it is worth noticing that by [29, Theorem 417C (v)], spt $\mu_{1,2}$ = spt $\mu_1 \times$ spt μ_2, and hence, if μ_1 and μ_2 are normalized in the sense that their support is included into $\mathcal{NF}(X)$, then spt $\mu_{1,2} \subseteq \mathcal{NF}(X)$ as well. Therefore, by Proposition 5.4 one might deduce that, if $\mathbf{b}_1$ and $\mathbf{b}_2$ are normalized belief functions, then $\mathbf{b}_{1,2}$ is normalized as well. The following example shows that this is not the case, since in the definition of $\mathbf{b}_{1,2}$, together with the product measure $\mu_{1,2}$, we also use the map $\rho^{\wedge}$ which, in fact, is not a genuine fuzzy-inclusion operator.

Example 7.3. Consider two belief functions $\mathbf{b}_1$ and $\mathbf{b}_2$ on $[0,1]^2$ with masses concentrated as follows:

$$\mu_1(1,0) = 1/4; \, \mu_1(1,1) = 3/4; \quad \mu_2(0,1) = 1/2; \, \mu_2(1,1) = 1/2.$$

Then, the product measure $\mu_{1,2}$ has support in the cartesian product of the supports of the two masses:

$$\{((1,0),(0,1)),((1,0),(1,1)),((1,1),(0,1)),((1,1),(1,1))\},$$

and it takes the following values:

$$\begin{aligned} \mu_{1,2}((1,0),(0,1)) &= 1/8, \\ \mu_{1,2}((1,0),(1,1)) &= 1/8, \\ \mu_{1,2}((1,1),(0,1)) &= 3/8, \\ \mu_{1,2}((1,1),(1,1)) &= 3/8. \end{aligned}$$

So, $\mu_{1,2}$ is normalized in the sense that each of its focal elements can be regarded as a normal fuzzy set in $[0,1]^2 \times [0,1]^2$. On the other hand, $\mathbf{b}_{1,2}$ is non-normalized: indeed, since $(0,0) = (1,0) \wedge (0,1)$, $\rho_{(0,0)}(0,0) = 1$ and $\rho_{(0,0)}(b,c) = 0$ for focal $(b,c) \neq (0,0)$, we have

$$\mathbf{b}_{1,2}(0,0) = \sum_{b,c:b \wedge c=(0,0)} \rho_{(0,0)}(b \wedge c)) \cdot \mu_{1,2}(b,c) = \mu_{1,2}((1,0),(0,1)) = 1/8 > 0.$$

The example we presented at the beginning of this section —the murder of Mrs Jones [52]— can be adapted to the context of belief functions on fuzzy sets as follows.

Example 7.4. Recall the 3 suspects of Mr. Jones' murder: Peter, Paul, and Mary. Consider the information provided by Mrs. Jones, she heard his husband yelling and the person she saw running was *a man* . It turns out that Mary has short hair, so she may be mistaken for a man at first sight, and hence the set of suspects looking like a man can be considered fuzzy as well, with membership function:

$$\mu_{man\text{-}like}(Peter) = 1, \ \mu_{man\text{-}like}(Paul) = 1, \ \mu_{man\text{-}like}(Mary) = 0.5.$$

The evidence supplied by Mrs Jones may be represented by a mass assignment $m_1 : [0,1]^X \to [0,1]$ such that $m_1(\textit{man-like}) = \alpha > 0$, $m_1(X) = 1 - \alpha$ and $m_1(a) = 0$ for any other $a \in [0,1]^X$.

A second piece of evidence is provided by the janitor living in the same house, who reports that he saw in the darkness a small person quickly leaving the scene of the crime. Paul and Mary are not tall while Peter is taller (Paul is 1.65 m tall, Mary is 1.60 m tall and Peter is 1.85 m). So, actually, the subset of small suspects of $X = \{Peter, Paul, Mary\}$ can be also considered as a fuzzy set, with membership function μ_{small} given by, say,

$$\mu_{small}(Peter) = 0, \ \mu_{small}(Paul) = 0.7, \ \mu_{small}(Mary) = 0.9.$$

The evidence supplied by the janitor may be represented by a second mass assignment $m_2 : [0,1]^X \to [0,1]$ such that $m_2(\textit{small}) = \beta > 0$, $m_2(X) = 1 - \beta$ and $m_2(a) = 0$ for any other $a \in [0,1]^X$.

Let us compute the resulting mass by combining m_1 and m_2 by means of the generalized Dempster rule according to Definition 7.2:

$$\mathbf{b}^{\wedge}_{1,2}(a) = \sum_{b,c \in \{man\text{-}like, small, X\}} \rho_a(b \wedge c) \cdot m_1(b) \cdot m_2(c) \ .$$

Here, the membership function of *small* $\wedge$ *man-like* (interpreting $\wedge$ by the min) is given by

$$\begin{aligned} \mu_{small \wedge man\text{-}like}(Peter) &= 0; \\ \mu_{small \wedge man\text{-}like}(Paul) &= 0.7; \\ \mu_{small \wedge man\text{-}like}(Mary) &= 0.5. \end{aligned}$$

Suppose we are interested in computing the belief that the suspect is Paul. We then need to compute:

$$\begin{aligned}\rho^{\wedge}_{\{Paul\}}(small \wedge man\text{-}like) &= \min_{x\in X}\{\mu_{small\wedge man\text{-}like}(x) \Rightarrow \mu_{Paul}(x)\} \\ &= \min\{0 \Rightarrow 0, 1 \Rightarrow 1, 0.5 \Rightarrow 0\} \\ &= \min\{1, 0.5\} = 0.5\end{aligned}$$

$$\begin{aligned}\rho^{\wedge}_{\{Paul\}}(small \wedge X) &= \min_{x\in X}\{\mu_{small}(x) \Rightarrow \mu_{Paul}(x)\} \\ &= \min\{0 \Rightarrow 0, 0.7 \Rightarrow 1, 0.9 \Rightarrow 0\} \\ &= \min\{1, 0.1\} = 0.1\end{aligned}$$

$$\begin{aligned}\rho^{\wedge}_{\{Paul\}}(X \wedge man\text{-}like) &= \min_{x\in X}\{\mu_{man\text{-}like}(x) \Rightarrow \mu_{Paul}(x)\} \\ &= \min\{1 \Rightarrow 0, 1 \Rightarrow 1, 0.5 \Rightarrow 0\} \\ &= \min\{0, 1, 0.5\} = 0\end{aligned}$$

and $\rho^{\wedge}_{\{Paul\}}(X) = 0$. Finally, we have

$$\begin{aligned}\mathbf{b}^{\wedge}_{1,2}(\{Paul\}) &= \sum_{a\in[0,1]^X} \rho^{\wedge}_{\{Paul\}}(b \wedge c) \cdot m_1(b) \cdot m_2(c) \\ &= \rho^{\wedge}_{\{Paul\}}(small \wedge man\text{-}like) \cdot m_1(man\text{-}like) \cdot m_2(small) + \\ &\quad \rho^{\wedge}_{\{Paul\}}(small) \cdot m_1(X) \cdot m_2(small) \\ &= 0.5 \cdot \alpha \cdot \beta + 0.1 \cdot (1-\alpha) \cdot \beta > 0.\end{aligned}$$

Hence, we get a positive belief degree of Paul being the murderer. This is in contrast with the results we would obtain, in case we assume Mary can be mistaken for a man, with both the classical model and the crisp-focal model, where focal elements are only allowed to be classical subsets of X. Indeed, in that case, we would be forced to take as focal element for m_1, besides X itself, the set *man-like* $= \{Paul, Mary\}$, and since there would be no focal element included in $\{Paul\}$, we would get $\mathbf{b}^{\wedge}_{1,2}(\{Paul\}) = 0$. □

The above min-conjunctive combination can easily be extended to well-known MV-operations on fuzzy sets, such as max-disjunction $\vee$, strong conjunction $\odot$ and strong disjunction $\oplus$, by defining

$$(b_1 \circledast b_2)(a) = s_{1,2}(\rho^{\circledast}_a),$$

for $\circledast$ being one of these operations, and defining

$$\rho^{\circledast}_a(b, c) = \rho_a(b \circledast c).$$

In this generalized case, the map $\mathbf{b}^{\circledast}_{1,2}$ resulting from the respective combination rule will be called the *$\circledast$-combination of* $\mathbf{b}_1$ *and* $\mathbf{b}_2$.

Whenever the supports of μ_1 and μ_2 are countable, it is easy to prove that $\mathbf{b}^{\circledast}_{1,2}$ is a belief function in the sense of Definition 4.9. In fact, in this case Definition 7.2 yields

$$\mathbf{b}^{\circledast}_{1,2}(a) = \sum_{b,c\in\mathbf{A}} \rho_a(b \circledast c) \cdot \mu_1(\{b\}) \cdot \mu_2(\{c\}). \tag{22}$$

Notice that (22) reduces to

$$\mathbf{b}^{\circledast}_{1,2}(a) = \sum_{d\in\mathbf{A}} \sum_{\substack{b,c\in\mathbf{A}\\ b\circledast c=d}} \rho_a(d) \cdot (\mu_1(\{b\}) \cdot \mu_2(\{c\})) = \sum_{d\in\mathbf{A}} \rho_a(d) \cdot \mu^*(\{d\}),$$

where

$$\mu^*(\{d\}) = \sum_{\substack{b,c\in\mathbf{A}\\ b\circledast c=d}} \mu_1(\{b\}) \cdot \mu_2(\{c\})$$

is indeed a mass assignment and hence $\mathbf{b}^{\circledast}_{1,2} \in Bel([0,1]^X)$. Therefore, turning back to the above Example 7.3 and Proposition 5.4, there exists a mass $\mu \neq \mu_{1,2}$ for $\mathbf{b}^{\circledast}_{1,2}$ such that spt $\mu \not\subseteq \mathcal{NF}(X)$.

8 Conclusions and Further Reading

In this chapter we have discussed several ways to define belief functions on MV-algebras of fuzzy sets based on some previous results by the authors [41,42,25,28]. In particular we have surveyed two main frames in which belief functions on fuzzy sets are characterized by the fact that focal elements are either crisp or fuzzy sets. We have then studied the normalization and soft-normalization problem together with a generalization of Dempster's rule of combination.

Another logical-based approach, extending the one in [30] for classical events, has been introduced in [27], where a modal logic for belief functions on an MV-algebra has been presented.

Belief functions on Boolean algebras can also be described in geometrical terms: in [9,10] the author presents several results in this directions. In a similar way, the problem of extending a partial assignment over formulas of Łukasiewicz logic can be characterized in geometrical terms by combining tropical-idempotent convex geometry and classical Euclidean convex geometry as well. The paper [22] studies this geometrical foundation for belief functions on MV-algebras.

The problem of extending a partial assignment to a probability measure is well known in the literature as de Finetti's coherence criterion [11,12]. As for belief functions on Boolean events, a similar interpretation of belief functions in terms of betting scheme has been presented in [38] and there is some ongoing work for the case of fuzzy events [23].

Acknowledgements. The authors would like to thank the anonymous referees for many valuable suggestions and remarks. The authors also acknowledge partial support by the FP7-PEOPLE-2009-IRSES project MaToMUVI (PIRSES-GA-2009-247584). Also, Flaminio acknowledges partial support of the Italian project FIRB 2010 (RBFR10DGUA-002), Kroupa has been supported by the grant GAČR 13-20012S, and Godo acknowledges partial support of the Spanish projects EdeTRI (TIN2012-39348-C02-01) and *Agreement Technologies* (CONSOLIDER CSD2007-0022, INGENIO 2010).

References

1. Belluce, L.P.: Semisimple Algebras of Infinite Valued Logic and Bold Fuzzy Set Theory. Canad. J. Math. 38(6), 1356–1379 (1986)
2. Biacino, L.: Fuzzy subsethood and belief functions of fuzzy events. Fuzzy Sets and Systems 158(1), 38–49 (2007)
3. Blok, W.J., Pigozzi, D.: Algebraizable logics. Memoirs of the American Mathematical Society 77(396) (1989)
4. Butnariu, D., Klement, E.P.: Triangular Norm Based Measures and Games with Fuzzy Coalitions. Kluwer, Dordrecht (1993)
5. Calvo, T., Mayor, G., Mesiar, R. (eds.): Aggregation Operators – New Trends and Applications. STUDFUZZ, vol. 97. Springer, Heidelberg (2002)
6. Chang, C.C.: Algebraic Analysis of Many-valued Logics. Trans. Am. Math. Soc. 88, 467–490 (1958)
7. Cignoli, R., D'Ottaviano, I.M.L., Mundici, D.: Algebraic Foundations of Many-valued Reasoning. Kluwer, Dordrecht (2000)
8. Cohn, P.M.: Universal Algebra. Revisited Edition. D. Reidel Pub. Co., Dordrecht (1981)
9. Cuzzolin, F.: A Geometric Approach to the Theory of Evidence. IEEE Transactions on Systems, Man, and Cybernetics, Part C 38(4), 522–534 (2008)
10. Cuzzolin, F.: The geometry of uncertainty. Information Science and Statistics Series. Springer (2012)
11. de Finetti, B.: Sul significato soggettivo della probabilità. Fundamenta Mathematicae 17, 298–329 (1931); Translated into English as "On the subjective meaning of probability". In: Monari, P., Cocchi, D. (eds.) Probabilità e Induzione, pp. 291–321. Clueb, Bologna (1993)
12. de Finetti, B.: Theory of Probability, vol. 1. Wiley, New York (1974)
13. Dempster, A.P.: Upper and lower probabilities induced by a multivalued mapping. The Annals of Mathematical Statistics 38(2), 325–339 (1967)
14. Denneberg, D.: Non-additive measure and integral. Theory and Decision Library. Series B: Mathematical and Statistical Methods, vol. 27. Kluwer Academic Publishers Group, Dordrecht (1994)
15. Denœux, T.: Reasoning with imprecise belief structures. Int. J. Approx. Reasoning 20(1), 79–111 (1999)
16. Denœux, T.: Modeling vague beliefs using fuzzy-valued belief structures. Fuzzy Sets and Systems 116(2), 167–199 (2000)
17. Develin, M., Sturmfels, B.: Tropical convexity. Doc. Math. 9, 1–27 (2004)
18. Dubois, D., Prade, H.: On several representations of an uncertain body of evidence, in. In: Gupta, M.M., Sanchez, E. (eds.) Fuzzy Information and Decision Processes, pp. 167–181. North- Holland, New York (1982)
19. Dubois, D., Prade, H.: Evidence Measures Based on Fuzzy Information. Automatica 21(5), 547–562 (1985)
20. Dubois, D., Prade, H. (eds.): Fundamentals of Fuzzy Sets. The Handbooks of Fuzzy Sets Series. Kluwer (2000)

21. Edwald, G.: Combinatorial Convexity and Algebraic Geometry. Springer, New York (1996)
22. Flaminio, T., Lacasa, L.G.: A note on the convex structure of uncertainty measures on MV-algebras. In: Kruse, R., Berthold, M., Moewes, C., Gil, M.A., Grzegorzewski, P., Hryniewicz, O., et al. (eds.) Synergies of Soft Computing and Statistics. AISC, vol. 190, pp. 73–82. Springer, Heidelberg (2013)
23. Flaminio, T., Godo, L.: Towards a betting interpretation for belief functions on MV-algebras. In: Mesiar, R., Pap, E., Klement, E.P. (eds.) Non-Classical Measures and Integrals: Abstracts of the 34th Linz Seminar on Fuzzy Set Theory (Linz 2013), pp. 51–53. Linz, Austria (2013)
24. Flaminio, T., Godo, L., Marchioni, E.: On the logical formalization of possibilistic counterpart of states over n-valued Łukasiewicz events. J. Logic Comput. 21(3), 447–464 (2011)
25. Flaminio, T., Godo, L., Marchioni, E.: Belief Functions on MV-Algebras of Fuzzy Events Based on Fuzzy Evidence. In: Liu, W. (ed.) ECSQARU 2011. LNCS, vol. 6717, pp. 628–639. Springer, Heidelberg (2011)
26. Flaminio, T., Godo, L., Marchioni, E.: Geometrical aspects of possibility measures on finite domain MV-clans. Soft Computing 16(11), 1863–1873 (2012)
27. Flaminio, T., Godo, L., Marchioni, E.: Logics for belief functions on MV-algebras. International Journal of Approximate Reasoning 54(4), 491–512 (2013)
28. Flaminio, T., Godo, L., Kroupa, T.: Combination and Soft-Normalization of Belief Functions on MV-Algebras. In: Torra, V., Narukawa, Y., López, B., Villaret, M. (eds.) MDAI 2012. LNCS, vol. 7647, pp. 23–34. Springer, Heidelberg (2012)
29. Fremlin, D.H.: Measure theory, vol. 4. Torres Fremlin, Colchester (2006); Topological measure spaces. Part I, II, Corrected Second Printing of the 2003 original
30. Godo, L., Hájek, P., Esteva, F.: A Fuzzy Modal Logic for Belief Functions. Fundamenta Informaticae 57(2-4), 127–146 (2003)
31. Goodearl, K.R.: Partially Ordered Abelian Group with Interpolation. AMS Math. Survey and Monographs 20 (1986)
32. Gottwald, S.: Fuzzy sets and fuzzy logic: the foundations of application–from a mathematical point of view, Teknea (1993)
33. Grabisch, M.: Belief Functions on Lattices. Int. J. Intell. Syst. 24(1), 76–95 (2009)
34. Grabisch, M., Murofushi, T., Sugeno, M.: Fuzzy Measure of Fuzzy Events Defined by Fuzzy Integrals. Fuzzy Sets and Systems 50, 293–313 (1992)
35. Hájek, P.: Metamathematics of fuzzy logic. Trends in Logic—Studia Logica Library, vol. 4. Kluwer Academic Publishers, Dordrecht (1998)
36. Halpern, J.H.: Reasoning about Uncertainty. MIT Press (2003)
37. Hwang, C., Yang, M.: Generalization of Belief and Plausibility Functions to Fuzzy Sets Based on the Sugeno Integral. International Journal of Intelligent Systems 22, 1215–1228 (2007)
38. Jaffray, J.-Y.: Coherent bets under partially resolving uncertainty and belief functions. Theory and Decision 26, 99–105 (1989)
39. Klir, G.J., Yuan, B.: Fuzzy Sets and Fuzzy Logic: Theory and Applications. Prentice Hall (1995)
40. Kroupa, T.: Every state on semisimple MV-algebra is integral. Fuzzy Sets and Systems 157, 2771–2782 (2006)
41. Kroupa, T.: From Probabilities to Belief Functions on MV-algebras. In: Borgelt, C., González-Rodríguez, G., Trutschnig, W., Lubiano, M.A., Gil, M.Á., Grzegorzewski, P., Hryniewicz, O. (eds.) Combining Soft Computing and Statistical Methods in Data Analysis. AISC, vol. 77, pp. 387–394. Springer, Heidelberg (2010)

42. Kroupa, T.: Extension of Belief Functions to Infinite-valued Events. Soft Computing 16(11), 1851–1861 (2012)
43. Litvinov, G.L.: Maslov Dequantization, Idempotent and Tropical Mathematics: a Brief Introduction. Journal of Mathematical Science 140(3), 426–444 (2007)
44. Mundici, D.: Averaging the truth-value in Łukasiewicz logic. Studia Logica 55(1), 113–127 (1995)
45. Mundici, D.: Advanced Łukasiewicz calculus and MV-algebras. In: Trends in Logic, vol. 35. Springer (2011)
46. Navara, M.: Triangular norms and measures of fuzzy sets. In: Klement, E.P., Mesiar, R. (eds.) Logical, Algebraic, Analytic, and Probabilistic Aspects of Triangular Norms, pp. 345–390. Elsevier (2005)
47. Panti, G.: Invariant measures on free MV-algebras. Communications in Algebra 36(8), 2849–2861 (2008)
48. Paris, J.B.: A note on the Dutch Book method, Revised version of a paper of the same title which appeared in. In: Proceedings of the Second Internat. Symp. on Imprecise Probabilities and their Applications, ISIPTA 2001, Ithaca, New York (2001)
49. Parsons, S.: Some qualitative approaches to applying the Dempster-Shafer theory. Information and Decision Technologies 19, 321–337 (1994)
50. Shafer, G.: A Mathematical Theory of Evidence. Princeton University Press, Princeton (1976)
51. Smets, P.: The Degree of Belief in a Fuzzy Event. Information Sciences 25, 1–19 (1981)
52. Smets, P.: Belief Functions. In: Smets, P., et al. (eds.) Nonstandard Logics for Automated Reasoning, pp. 253–277. Academic Press, London (1988)
53. Sugeno, M.: Theory of fuzzy integrals and its applications. Phd. Dissertation, Tokyo Institute of Technology, Tokyo, Japan (1974)
54. Weber, S.: $\perp$-decomposable measures integrals for Archimedean t-conorms $\perp$. J. Math. Anal. Appl. 101, 114–138 (1984)
55. Yen, J.: Generalizing the Dempster-Shafer theory to fuzzy sets. IEEE Transactions on Systems, Man, and Cybernetics 20(3), 559–570 (1990)
56. Yen, J.: Computing Generalized Belief Functions for Continuous Fuzzy Sets. Int. J. Approx. Reasoning 6, 1–31 (1992)
57. Zadeh, L.A.: Fuzzy sets. Information and Control 8, 338–353 (1968)
58. Zadeh, L.A.: Probability measures of fuzzy events. J. Math. Anal. Appl. 23, 421–427 (1968)
59. Zadeh, L.A.: Fuzzy sets and information granularity. In: Gupta, M., et al. (eds.) Advances in Fuzzy Sets Theory and Applications, pp. 3–18. North Holland (1979)
60. Zimmermann, H.J.: Fuzzy sets in operational research. European Journal of Operational Research 13, 201–216 (1983)
61. Zhou, C.: Belief functions on Distributive Lattices. In: Proceedings of the Twenty-Sixth AAAI Conference on Artificial Intelligence, pp. 1968–1974 (2012)

Author Index

Printed in the United States
By Bookmasters